HARDY'S METRES AND VICTORIAN PROSODY

Thomas Hardy, 1910, by William Strang. By permission of the
Fitzwilliam Museum, Cambridge.

HARDY'S METRES
AND
VICTORIAN PROSODY

With
A Metrical Appendix of Hardy's Stanza Forms

DENNIS TAYLOR

CLARENDON PRESS · OXFORD
1988

Oxford University Press, Walton Street, Oxford OX2 6DP

Oxford New York Toronto
Delhi Bombay Calcutta Madras Karachi
Petaling Jaya Singapore Hong Kong Tokyo
Nairobi Dar es Salaam Cape Town
Melbourne Auckland

and associated companies in
Berlin Ibadan

Oxford is a trade mark of Oxford University Press

Published in the United States
by Oxford University Press, New York

British Library Cataloguing in Publication Data
Taylor, Dennis, 1940–
Hardy's metres and Victorian prosody:
with a metrical appendix of Hardy's
stanza forms.
1. Hardy, Thomas, 1840–1928. Poetic works
I. Title
821'.8   PR4557.P58
ISBN 0–19–812967–X

Library of Congress Cataloging in Publication Data
Taylor, Dennis, 1940–
Hardy's metres and Victorian prosody.
Bibliography: p.   Includes index.
1. Hardy, Thomas, 1840–1923—Versification.
2. English language—Versification.   I. Title.
II. Title: Hardy's meters and Victorian prosody.
PR4757.V4T39   1988   821'.8   87-31241
ISBN 0–19–812967–X

Typeset by Eta Services (Typesetters) Ltd.

Printed in Great Britain
at the University Printing House, Oxford
by David Stanford
Printer to the University

To My Parents

Simplicity assuages
With grace the damaged heart,
So would I in these pages
If will were art.

But the best engineer
Of metre, rhyme, and thought
Can only tool each gear
To what he sought

If chance with craft combines
In the predestined space
To lend his damaged lines
Redeeming grace.

J. V. Cunningham, 'The Predestined Space'

Acknowledgements

A Boston College Mellon Grant enabled me to complete the final portions of this manuscript. Since this book began as an extensive chapter in an earlier manuscript eventually condensed and published as *Hardy's Poetry, 1860–1928* (London, 1981), the acknowledgements made there should also be assumed here. I am particularly grateful to Terry Brogan, Andrew Von Hendy, Samuel Hynes, and Edwin Pritchard, who read complete drafts and made many helpful suggestions. I am also indebted, for helpful correspondence or conversation, to James Gibson, David Holmes, Richard Purdy, S. J. Keyser, Anne Ferry. Boston College and its English department have been for me a stimulating and supportive environment, as was, for a briefer period, the Thomas Hardy Society Conference of 1986 at Dorchester. I am grateful to Roger Peers at the Dorset County Museum, for delivering to me several armloads of Hardy books. Hardy made a great number of interesting annotations in these books, only a portion of which have been noted in this book, in Wright, and in Björk (Hardy, *Literary Notebooks*). My wife, Mary, and my children, John, Matthew, Kathryn, and Mary Rebecca, have continued to teach me more than I could ever learn from books. Finally, this book is dedicated with the old affection to my parents, Mr and Mrs Frank E. Taylor of Baltimore, Maryland.

Contents

Note on References and Abbreviations

Through the text and notes references are made, by author name and volume/page number, to works cited in the bibliography; in the case of more than one work by the same author appearing in the bibliography a short title has been added to references to avoid confusion. However, for repeated citations the following abbreviated forms have been used. For full bibliographical details, see the bibliography. Works whose full details are given in text or notes do not usually appear in the bibliography.

Bridges, *Milton's Prosody*	Bridges, Robert, *Milton's Prosody* (Oxford, 1921).
Collected Letters	*The Collected Letters of Thomas Hardy*, ed. Richard Purdy and Michael Millgate, 6 vols. (Oxford, 1978–87).
Complete Poetical Works	*The Complete Poetical Works of Thomas Hardy*, ed. Samuel Hynes, 3 vols. (Oxford, 1982–5).
De Selincourt	De Selincourt, Basil, 'English Prosody', *Quarterly Review*, 215 (July, 1911), 68–96.
Heine	Heine, Heinrich, *Poems*, trans. Edgar Alfred Bowring (London, 1878).
Henley	Henley, W. E. (ed.), *English Lyrics: Chaucer to Poe 1340–1809* (London, 1897).
Hullah	Hullah, John (ed.), *The Song Book: Words and Tunes from the Best Poets and Musicians* (London, 1866).
Hymns	*Hymns Ancient and Modern* (1860–1).
Life	*The Life of Thomas Hardy 1840–1928 by Florence Emily Hardy* (London, 1962).
Linton	Linton, W. J. and R. H. Stoddard (eds.), *English Verse: Lyrics of the 19th Century* (London, 1884).

Literary Notebooks	*The Literary Notebooks of Thomas Hardy*, ed. Lennart Björk, 2 vols. (New York, 1985).
Omond, *English Metrists*	Omond, Thomas, *English Metrists* (Oxford, 1921).
Palgrave	Palgrave, F. T., *The Golden Treasury* (Cambridge, 1861).
Patmore, *Essay*	Patmore, Coventry, *Essay on English Metrical Law*, ed. Sister Mary Roth (Washington, DC, 1961).
Personal Notebooks	*The Personal Notebooks of Thomas Hardy*, ed. Richard H. Taylor (London, 1979).
Personal Writings	*Thomas Hardy's Personal Writings*, ed. Harold Orel (Lawrence, 1966).
Saintsbury	Saintsbury, George, *A History of English Prosody*, 3 vols. (London, 1906–10).
Schipper	Schipper, Jacob, *Englische Metrik: in Historischer und Systematischer Entwickelung Dargestellt*, 2 parts (Bonn, 1881–8).
Taylor, *Hardy's Poetry*	Taylor, Dennis, *Hardy's Poetry, 1860–1928* (London, 1981).
Wright	Wright, Walter, *The Shaping of 'The Dynasts'* (Lincoln, Nebr., 1967)

The bibliography contains the following abbreviations:

ELH	*English Literary History*
JAAC	*Journal of Aesthetics and Art Criticism*
JEGP	*Journal of English and Germanic Philology*
MLN	*Modern Language Notes*
MLQ	*Modern Language Quarterly*
MLR	*Modern Language Review*
PMLA	*Publications of the Modern Language Association*
TLS	*Times Literary Supplement*

In metrical analyses I have used combinations of the following abbreviations:

d.	duple
f.	falling
r.	rising
t.	triple

Introduction

WHEN Hardy turned from writing novels to writing poems, his reviewers were not very happy. They were rarely satisfied with Hardy's claim that he was primarily a poet and only secondarily a writer of prose. In 1902 Hardy said he had returned to poetry 'because that form of expression seems to fit my thoughts better as I grow older, as it did when I was young also'; in 1904 he said he 'found the condensed expression that it affords so much more consonant to my natural way of thinking & feeling'. In 1911 Hardy wrote a preface for the Wessex Edition of his works and reviewed his career:

The few volumes filled by the verse cover a producing period of some eighteen years first and last, while the seventeen or more volumes of novels represent correspondingly about four-and-twenty years. One is reminded by this disproportion in time and result how much more concise and quintessential expression becomes when given in rhythmic form than when shaped in the language of prose.[1]

In 1912 Hardy advised a friend to read the poems rather than the novels because 'the novels seem immature to me'. In 1915, he told Harold Child 'to treat my verse . . . as my *essential* writings, & my prose as my *accidental*'. The verse was 'the key' to his novels, he said in 1916. Objecting in 1918 to a critic who claimed 'Hardy is a realistic novelist who . . . has a grim determination to go down to posterity wearing the laurels of a poet', Hardy noted: 'At the risk of ruining all my worldly

---

[1] The eighteen-year span Hardy cites includes the sixteen years, 1896–1911, when he was preparing *Wessex Poems*, *Poems of the Past and the Present*, *Time's Laughingstocks*, and *The Dynasts*, also the early years when he began writing poetry, especially 'the years 1866 and most of 1867' when he 'wrote constantly' and the balance constituted by those months which Hardy set aside for poetry during his architectural and novel-writing years.

prospects I dabbled in it [poetry] . . . was forced out of it. . . . It came back upon me.' When given a copy of the 1920 edition of Louis Untermeyer's *Modern British Poetry*, he 'revised the text (in pencil), deliberately promoting his reputation as a poet rather than a novelist'. In 1922, noting again the number of his poetry-writing years, he said: 'This disposes of the cuckoo-cry that T.H. being new to poetry expresses himself clumsily in verse, his verse . . . being, as his long practice would suggest, and the best critics acknowledge, more finished than his prose, except where intentionally rough hewn.' In the same year, in the manuscript version of his 'Apology' to *Late Lyrics and Earlier*, he objected to being labelled a novice at poetry 'when by reference to dates it could be seen that he has written verse thirty years, and prose about twentyfive at the outside'. In 1923, he claimed that his novels 'have been superseded in the view of critics by the more important half of my work, the verse, published during the last 25 years'. When he died in 1928, he had been a full-time poet for thirty-six years.[2]

Why was Hardy so interested in metrical forms? By my count he composed approximately 1,093 poems. These include eight volumes of poetry, uncollected verse, and distinguishable verse forms used in *The Dynasts* and *The Queen of Cornwall*.[3] What did Hardy find so fascinating about the act of putting his language into a structure of rhymes and patterned rhythms? For what purposes did he use the various forms he chose? Who influenced him?

Hardy's poetic career lasted from 1860 to 1928, thus encompassing the mid- and late Victorian periods, the Edwardian and Georgian periods, the post-war period, and the

[2] *Collected Letters*, iii. 43, 133; *Personal Writings*, 48; *Collected Letters*, iv. 220; v. 94; v. 172; *Life*, 415; Holmes, *English First Editions*, item 96; *Collected Letters*, vi. 157; *Complete Poetical Works*, ii. 323; *Collected Letters*, vi. 182; see also *Life*, 48, 57, 100, 179; *Collected Letters*, v. 76, 253. In 1922, Hardy reminded a critic that in the 1890s he 'turned back to poetry' (where the critic had written 'turned to poetry') (*Collected Letters*, vi. 156). References to Hardy poems, by volume and page, are to the *Complete Poetical Works*, ed. Samuel Hynes (Oxford, 1982–5). In the case of variant readings, I have also consulted the *Complete Poems*, Variorum edn., ed. James Gibson (London, 1979).

[3] The total, I estimate, includes: 918 poems in the eight collected volumes; 26 uncollected poems; 113 distinguishable verse forms in *The Dynasts*; 36 verse forms in *The Queen of Cornwall*.

twenties. This time-span represents the climax and end of the 500-year era of accentual-syllabic verse in English, that is the era when rhymed or blank accentual-syllabic verse was the norm, not the exception. As the accentual-syllabic era recedes, it becomes more difficult for the modern reader to enter into the spirit of Hardy's techniques and forms. Our understanding and appreciation of these forms is becoming more perplexed. Indeed, we may soon be in the situation of those medieval Latinists whose understanding of quantitative rhythms became more perplexed, as the new accentual-syllabic rhythms took over.

Why in the Victorian period which first nourished Hardy was there such an extraordinary range of metrical poetry? Why was this the last of the great periods of traditional metrical poetry? What was the Victorian conception of the nature and function of metrical form? What was Hardy's relation to Victorian theory and practice?

These questions about Hardy's metres and about Victorian metrical theory eventually lead (or led me) into a yet larger question: what is the nature of English metre? How has it been understood over the centuries, and how did the Victorian metrical theorists understand it? In my own research, these three topics kept connecting with each other: the nature of metre, the Victorian conception of the nature of metre, Hardy's conception of the nature of metre. It eventually seemed clear to me that Hardy's fascination with metrical form was a fascination born and nurtured within the context of Victorian metrical theory and practice; and that together Hardy and the Victorian theorists have much to offer us in the way of understanding metre itself.

I have emphasized Victorian theory for a number of reasons. It is in the theorists that we can find articulated much of what inspired Victorian metres; in fact, many of the poets were theorists and their practice and their theory went hand in hand. This complementarity is more true of the Victorian poets than of earlier poets whose practice was not in fact much illumined by the reigning metrical theory. In fact, the Victorian period was the first period to discover a theory of metre adequate to the genius of its poets. By contrast, for example, Renaissance classical theory was inadequate for the poetry of Sidney, Donne, and Shakespeare. But Tennyson, Browning, and Swinburne had some good critics associated with them or closely following upon them. This is the first argument I make in this book.

The argument is peculiarly appropriate for a study of Hardy, because of his immense and not entirely appreciated importance as a Victorian metrical poet. His career as a metrical poet was longer than that of any other Victorian poet, indeed any English poet. He wrote poems in more metrical forms than any other major English poet, indeed perhaps than any other poet. Hardy's life-long commitment was to stanzaic forms. There is very little blank verse in the *Complete Poetical Works*, and no free verse. All the while, he was an avid reader of the works of his contemporaries. His extensive notebooks are still a largely untapped mine of quotations from hundreds of Victorian writers on every subject, including metrical subjects. He was acutely responsive, sometimes excessively so, to the currents of his age. His conception of metrical form matured over his lifetime, a maturity that can be seen in the way he used metre in his poems. The more he wrote poetry, the more he read in the metrical theorists. His career, from 1860, the date of his earliest known poem, to 1928, the date of his last poem, matches the climactic period of accentual-syllabic theory, from Patmore's landmark article of 1857 to the beginnings of the new criticism.

I do not mean to suggest that Hardy read the theorists and then went off and wrote his poems. Rather he and they participated in a common exploration, a community of metrical discovery which inspired theory and practice.

Chapter 1 presents a history of metrical theory and its distinctive resolution in the Victorian period. I have taken some pains with details of this history, in order to show why the Victorian resolution *was* distinctive. This history is such as Hardy might have seen it. But the history may also stand alone and be of interest to readers interested in general questions of metrical theory. Chapter 2 discusses in some detail the relation of Hardy to the tradition of poetry and theory, after he arrived in London in the 1860s. Chapter 3 sets up some terms by which we can appreciate the skill of Hardy's metrical techniques. Chapter 4 uses these terms to study metrical development in a series of Hardy's poems. Chapter 5 suggests some final reflections on Hardy's relation to the tradition of visual stanzas in English poetry.

If I can sum up the book's overall argument in a few words, I would say that what the Victorian period achieved in metrical theory was an understanding of the abstract nature of metrical form and the dialectic way in which it interplayed with the spoken language. This may sound familiar and indeed it is the central insight of twentieth-century new critical studies of metrical forms. The Victorian sources of this insight have, however, been largely ignored. Second, I would suggest that it was the dynamic nature of this abstraction which so intrigued Hardy that it led him to his three-score-year pursuit of the subject. When he arrived in London in the 1860s, he found himself surrounded by an extraordinary efflorescence of sonnet forms, hymn forms, ballad forms, classical forms, romance forms, and mainstream lyric poetry. At the same time, the influence of Patmore's extraordinary article was beginning to make itself felt. Interestingly, one of the things that drew Hardy close to the interests of the metrical theorists was the aesthetic of Gothic architecture. This aesthetic played an important role in the new metrical theory; and Hardy, as one of the last of the Gothic Revival architects, was peculiarly positioned to appreciate its force. Theory and practice, then, influenced Hardy as he began his career with sonnets, ballads, and classical experiments.

Hardy's poems represent not only a dazzling array of metrical forms but an equally rich resource of metrical effects. Chapter 2 discusses the technical constituents of Hardy's verse, defines them in a simple way, in order to see how they are combined and recombined in more complex ways in these series of imitations. Also described here is Patmore's important discovery of the dipodic principle, and Hardy's rich adaptation of the principle half a century later in 'After a Journey'. In Chapter 3, I have focused on two kinds of metrical effect which Hardy perfects. One is where the metre imitates various kinds of interactions of the mind with the physical world. The other metrical effect is where the metre serves as an abstract form which represents a final stage of thinking and speaking. These two kinds of metrical effect enable us to trace in Chapter 4 a pattern of metrical development in many of Hardy's poems. Several series of Hardy's poems imitate with increasing complexity the ways in which the mind interacts with the world and the way in which this interaction 'evolves' into a fixed metrical form. The mind may interact with rhythms of the clock, or the rhythms of music, or of a storm, or of light. The place of *The Dynasts* in Hardy's metrical development is also briefly considered.

Chapter 5 focuses on the idea of the stanza form, both as an interesting technical entity and as a mimetic vehicle. Hardy's acute sense of the stanza's visual shape, rooted in inscription poetry, reflects the visual analogies latent in traditional discussions of accentual-syllabic metre. Here, we shall consider Hardy's last and most unusual metrical experiments, poems written in unique, complex stanzas. These experiments connect Hardy's poetry with the modern development of free verse. Indeed, through Hardy's eyes we can see an intimate connection between free verse and traditional accentual-syllabic verse. Hardy is an important transitional figure in this historical development.

This book includes a Metrical Appendix which indicates the parallels between Hardy's metrical forms and the tradition of metrical forms which he may have known. The richness of Hardy's relation to the metrical tradition is perhaps most palpable in the Appendix. The Appendix will itself stand as an extensive compendium of metrical forms in English. Whatever its value in this era of free verse, it is a tribute to the breadth of Hardy's metrical experimentation.

W. H. Auden said that Hardy was his 'first master': 'his metrical variety, his fondness for complicated stanza forms, were an invaluable training in the craft of making' (*Making*, 38). This book tries to throw some light on the sense of that remark; but first we must plunge into the byways of metrical history, and chart the route that leads to Hardy.

1. *Victorian Prosody and its Backgrounds*

ONE of the puzzles of English literary history is why it took so long to evolve an adequate description of standard English metrical rhythm—the rhythm we associate with the major tradition from Chaucer to Tennyson. In 500 years there had been some advances but not many. 'The history of criticism', Hardy said after a discussion of Barnes's language and rhythm, 'is mainly the history of error, which has not even, as many errors have, quaintness enough to make it interesting' (*Personal Writings*, 81). English metrical rhythm is a phenomenon governed by laws which are extremely hard to define though we experience them clearly enough. If metrical rhythm is a measured recurrence of something in the language, what is that something and how is it measured? If we think we have it—say, a recurrent accent on certain numbered syllables—how do we reconcile apparent variation with that measure? The centuries old puzzle of English rhythm is that once a metrical theorist finds a way to formulate the metrical rule of a given poem, a reader will soon find a line which violates the formulation. Indeed a wide variation in number and placement of accents often seems consistent with the basic rhythm. How then can we define a law which is general enough to cover these variations and yet strict enough to serve as an explanation—so that, for example, one could distinguish metrical from unmetrical lines? When the metrical art of verse depends on a distinctive rhythm, and this rhythm depends on an organization of syllables and stresses, it remains puzzling why this aspect of the art cannot be successfully described. Understanding metrical law is important because metrical law enables us to read poetry as the poet intended it—often with the right emphasis, the right phrasal structure, and with the right sense of the artful relation of sound and sense, or at least with the right sense of the range of possibilities of these things. For readers who lack an instinctive awareness of a poem's rhythm, or for readers faced with an older form of poetry, instruction in metrical law may be necessary simply to learn to read correctly.

To be puzzled at the problem is often the first step toward appreciating metre. T. S. Omond, the leading metrical historian up to T. V. F. Brogan, says of William Mitford, one of his major figures: 'Most valuable of all, perhaps, was his explicit declaration that English verse-structure presents a problem to be solved.' When accentual-syllabic rhythm emerged in medieval Latin verse and became popular in church hymns, the theoreticians of this new rhythm were likely to be puzzled. The origin of the phrase, 'without rhyme or reason', may derive indirectly from the medieval contrast (in Bede for example) between the *ratio* of Latin metre with its explicit laws of quantity and the lack of *ratio* in the new 'rhythmus', i.e. 'rhythmus sine ratione', 'rime without reason'. In the 1860s Hardy read, in Henry Hallam's *Introduction to the Literature of Europe in the Fifteenth, Sixteenth, and Seventeenth Centuries* (1837–9), Hallam's summary of the way accentual metre emerged in Latin: 'All metre, as Augustin says, was rhythm, but all rhythm was not metre: in rhythmical verse, neither the quantity of syllables . . . nor even, in some degree, their number, was regarded, so long as a cadence was retained in which the ear could recognise a certain approach to uniformity.' Our standard English rhythm, with its complicated heritage in 'French and Anglo-Saxon principles . . . with music, religious and secular, and Latin for intermediaries' (as Hardy read in De Selincourt) has remained a rich and puzzling phenomenon.[1]

This basic contrast between quantitative metre and the new *rhythmus*—based somehow on accents and syllables—is behind the various words and phrases used to describe our standard rhythm: 'accentual-syllabic', 'syllable-stress', 'syllable-accent',

[1] Omond, *English Metrists*, 79; Hallam, i. 39; De Selincourt, 85. Hallam's book is cited in *Desperate Remedies*, 6.3. On 'rhythmus sine ratione' see J. W. Rankin, 'Rime and Reason'.

or merely 'accentual'. 'Accentual' and 'accentual-syllabic' remained synonymous terms until the twentieth century.[2]

The History of English Prosody

The problem of the new rhythm 'sine ratione' has still not been entirely solved and continues as the subject of much discussion. Nevertheless there have been some advances in the half-millenium history of English metrical theory—and these advances have occurred in several discernible stages, roughly chronological, though some explanations devised in earlier stages have remained in use long after, and parts of later explanations can be found earlier. In each stage the metrical law is defined in a distinctive way—as classical law, as mechanical law, as musical law, as organic law, as dialectical law, as statistical law, as structural law, as generative law. Such definitions tend to reflect the character of the age in which they were proposed: the classical analogy promoted in Renaissance England and influential thereafter, the mechanical theory sketched in the Renaissance but dominant in the Age of Reason, the musical theory developed in the late eighteenth century, the organic theory in the Romantic period, the dialectical theory or 'new prosody' established in the Victorian period, the statistical theory becoming prominent at the beginning of the twentieth century, followed eventually by structuralism and then the era of Chomsky. Metrical theory seems peculiarly dependent on analogies and world views because the subject itself is so resistant to simple analysis.

The classical theory defined the English line by analogy with the Greek and Roman line, as a sequence of long and short syllables organized into feet. The feet, according to one version

---

[2] See *OED*. This historically standard definition of the word 'accentual' needs to be distinguished from 20th-century notions of 'accentual' or what I would call 'accentualist' metre, supposedly based merely on the number of accents per line and associated with Old English verse. Bridges, for reasons of his own, uses the term 'syllabic' to refer to the accentual-syllabic tradition (see below). In *Three Models for English Verse*, T. V. F. Brogan traces the confusions in metrical history resulting from prosodists trying to manage three different ways of calculating a line's metre, by number of syllables, by number of accents, and by number and type of feet.

of the theory, were equivalent in quantity of duration so that, for example, two shorts would equal one long.[3] According to this theory, the syllables which filled the feet could be adjusted in certain ways, provided the proportions were observed: a spondee (--) could be substituted for a dactyl (-ᵥᵥ) in the dactylic hexameter.

In theory this made sense but in practice did not work very well, even as a description of classical Greek and Latin verse. In classical verse we find that in some metres substitutions of certain feet, though mathematically possible, are forbidden by tradition; in other metres substitutions of certain feet, though mathematically faulty, are allowed. Many verse forms seem organized not according to feet but according to larger groupings of feet: dipodies and cola. Moreover, some theorists argue that quantities fill up rather than determine the intervals constituted by the rhythmic beat. It is also unclear whether syllables were 'long' because they were long in actual duration, or because they were declared long by convention—in which case, in turn, they might come to seem long to the ear. The role of stress in the classical metres, especially in Latin metres, is also unclear. According to various theories and depending on the various stages of Latin history and Latin metrical theory, Latin stresses seem to support or conflict with or be indifferent to Latin quantities.

The application of classical concepts to English metre was particularly confusing because English syllables do not have clearly defined quantities, and stress more than quantity determines the rhythm of the English line. Thus quantities cannot provide the ratios necessary for the English metrical pattern. Quantities do influence stresses; they affect for example the redistribution of stresses in polysyllabic words. (We change the noun 'person' to the adjective 'personal', but 'parent' to 'parental'.) And quantities have some effect, as yet not clearly formulated, on the English metrical line. But stresses can easily override quantities. Spenser noted this fact when he

---

[3] '. . . all syllables being presumed to fall under a known division of long and short, the former passing for strictly the double of the latter in quantity of time'—as Hardy read in Hallam's summary (Hallam, i. 38–9).

called the quantitative pronunciation of 'carpenter'—
carpénter—a 'lame gosling that draweth one leg after her'. I
believe Ransom is adverting to this famous observation in his
poem 'Captain Carpenter'. We would like to pronounce the
poem's opening line 'Cáptain Carpénter rose úp in his príme',
but we can't.

Thus the classical law, muddled in itself and applied to
English verse, was hardly a law at all. It was a tradition that
the units of English rhythm were somehow comparable to the
units of classical rhythm, and that substitution of different feet
should be like classical substitutions. Our earliest Renaissance
prosodies struggled with this issue, invoking the classical
analogy, yet also trying to distinguish English verse by its
rhymes, number of syllables, and number and position of
stresses. The classical analogy enjoyed its greatest *éclat* in
imitations and discussions by Sidney, Spenser, Campion, and
others in the sixteenth and seventeenth centuries. The mechanical theory would later formalize this sort of numeration. For
four hundred years the use of 'feet' would remain a good way of
picturing a line's metrical form, but could not explain its law.
'Feet' could not explain the difference between a violation and
a variation of that form.

The Age of Reason demanded a real law, as powerful as
those governing phenomena in the physical sciences. Such a
law specified the number of syllables and the placement of
stresses in an English line. The eighteenth century thus affixed
a mathematical aura to the term 'numbers', which had
traditionally denoted proportion and harmony. The law of
iambic pentameter, then, prescribed a succession of ten
syllables in which an accent appeared on even syllables—
neatly corresponding to five iambs on the classical model.
These two models were easily interchanged,[4] and apparent
variations could be ironed out through devices like elision.[5]

[4] Charles Gildon says that critics like Bysshe 'erroneously use *Accent* for *Quantity*'; but
Gildon's quantities are not really quantities but in effect what Bysshe calls accents. See
Gildon's 'New Prosodia', from John Brightland's *A Grammar of the English Tongue*
(1711), 153. This chapter should be attributed to Gildon, *contra* Omond, 33, *pro* Culler,
'Edward Bysshe', 878, citing Herman Flasdieck's discovery.

[5] Paul Fussell minimizes unduly the prescription for accent placement in 18th-

The trouble with this mechanical law was that it produced monotony if carried out to the letter, and in any event was an inadequate description of the sinewy grace of Renaissance poetry or even the metrical subtlety of the best eighteenth-century verse. So other strategies were tried. Dr Johnson, typically, allowed that variation relieves monotony but also 'makes us more sensible of the harmony of the pure measure' (*Rambler*, 86). Variety was treated as a poor relative, and its behaviour was narrowly confined and sometimes restricted to mere play with caesura placement, as in Edward Bysshe. As the eighteenth century advanced, the expressive value of variety was more and more defended but its theoretical justification remained slight. Tyrwhitt, regularizing Chaucer in 1775, noted: 'It is agreed, I believe, that in our heroick metre those verses (considered singly) are the most harmonious in which the accents fall upon the even syllables; but it has never (that I know) been defined how far a verse may vary from this its most perfect form, and yet remain a verse.' Lord Kames tried to divide up the 'principle of variety in uniformity' in a typically mechanical or Bysshean way: 'uniformity prevails in the arrangement, in the equality of the lines, and in the resemblance of the final sounds; variety is still more conspicuous in the pauses and in the accents, which are diversified in a surprising manner'. How, and how far, uniformity and variety could apply to the same elements was still a riddle waiting to be solved. Frances Hutcheson, who influenced Samuel Say, celebrated '*uniformity amidst variety*' as though each could be clearly confined.[6]

The mechanical theory was guilty of what Whitehead later called the fallacy of misplaced concreteness. Though sometimes

century theory (*Theory*, 14–15). Fussell follows Dwight Culler's more careful discussion in 'Edward Bysshe and the Poet's Handbook'. Culler discovered that Bysshe had modelled his treatise on Claude Lancelot's 'Breve Instruction sur les Regles de la Poësie Françoise'. But it is not quite correct to say with Culler that Bysshe's ' "Rules" are simply a translation and adaptation of the "Breve Instruction" with English examples replacing the French' (877). It is true that Bysshe felt he could lift the syllabic discussion out of Lancelot; but onto this he intrudes considerations of Accent.

[6] Tyrwhitt, sect. xvii, footnote, later cited by Bridges, *Milton's Prosody*, 84; Kames, ch. 18, sect. iv; Hutcheson, 357.

seeming to rely on a Platonic ideal, such theorizing was in effect a simple empirical description of a standard line which was then made to exclude other kinds of lines. The important distinction between two orders of discourse, talk about the metre and talk about a given line's rhythm, was not yet clearly made. The mechanical theory specified too concretely and directly how the syllables of a given line should be numbered and accented. The relation of metrical law to poetic line was not yet properly conceived.

What we see in the next stages is a gradual achievement in placing the law of metre at the right level of abstraction. Metrical law, so the nineteenth century discovered, does specify number and accent-placement of syllables, but on a level not identical with any particular manifestation. The emergence of this sense of abstract form is a slow and intermittent process. Many things contributed, from the new philosophical climate associated with Kant and Hegel to changes in traditions of poetic recitation, that is the growing tendency not to pronounce those elisions but still keep their punctuation as fictions referring to a metrical structure (see Culler, 884). The nominalistic and Hobbesian tendencies of eighteenth-century empiricism tended to obscure the notion of a powerful abstraction. For example, when William Mitford in 1774 says that in spite of some variations—for example a trochee for an iamb at some point in a pentameter line—the line retains its 'fundamental form' (100), he means that the other feet are iambs. Peter Fogg, late in the century (1792–6), is ready to admit substitutions 'but under considerable restrictions'. An anapaest, Fogg says, may be substituted for an iamb in an iambic pentameter 'provided the greater part of the line remain iambics' (ii. 189). 'The pyrrhic ought to be compensated by a spondee posterior to it, which reduces the quantity of the two to an exact equivalence with two iambics.' This is a more exact mathematical notion than the notion of substitution which Saintsbury, for example, will later invoke. Only gradually is it conceived that a fundamental form can remain though many more of the 'feet' are variations of the norm, and indeed that the metrical pattern controls the poetic line in a more complex way than had been previously discerned.

Unfortunately, both classical theory and mechanical law theory have persisted to the present day, much to the dismay of poets who were developing the full possibilities of the accentual-syllabic tradition.[7]

The musical theorists represent an important stage in this growing conception of abstract metrical form. The most prominent names before the twentieth century are Joshua Steele, Richard Roe, and Sidney Lanier. Indeed, the images available for prosody were influenced by the developments taking place in musical notation since the middle of the seventeenth century, namely the increasingly uniform use of the 'isochronous' bar-line. The bar in a vivid visual way revealed a regularity of spacing in its box-like enclosure even though each box was filled with a differing number and type of marks denoting the duration and scale of individual musical notes. This clear picture of variety within uniformity could be neatly applied to English rhythm. The poetic line could be divided into bars and a varying number of syllables and stresses (or pauses) could fill each bar. Hardy was influenced by this tradition when he said in 1904 that in poetry 'rhythm and rhyme are a non-necessitous presentation of language under conditions that in strictness appertain only to music' (*Personal Writings*, 143). The problem was: if in poetry neither the syllables nor the stresses indicated the bar divisions, what did indicate them? Poetry usually did not have the benefit of a supplementary musical accompaniment. Moreover the same words can be set to different time signatures. In poetry the syllables and stresses had to govern themselves. Thus while the musical theorists introduced an important image into the

---

[7] Omond would later call these traditions the 'Old Orthodoxy' (*English Metrists*, 1–6); Saintsbury said they represented the 'preceptist' approach to prosody (ii. 200; iii. 151; ii. 541). Each of these traditions can be heard in the many negative reviews of Hardy's verse. The mechanical theorists were valuable, however, in posing a challenge to later metrists: produce a better law which is truly a law. Saintsbury's grudging praise of Bysshe should still be read (ii. 541). Saintsbury has good remarks on the blending of the syllabic and foot theories of prosody: see *Historical Manual*, 19; see also Culler, 'Edward Bysshe', 878–9.

discussion,[8] they by no means solved the problem. In the history of metrics, they have tended to remain a minority party.

The problem remained for the nineteenth century: how can the form remain if the substance changes? How is the metre to be defined if the material elements of the metre vary? Variety had been kept within fairly tight bounds by the two earliest theories; either it had been forcefully reconciled with law, or grudgingly allowed for expressiveness, or condemned for going too far. What intrigued the nineteenth century was the legal status of variety, what Southey in the preface to *A Vision of Judgment* called 'the union of variety with regularity', or what Leigh Hunt described (much to Poe's irritation) as the 'principle of Variety in Uniformity'. Such critics were not so neatly dividing the issue as did Hutcheson earlier.

The Romantic reaction against eighteenth century mechanism took the form of a defence of variety for its own sake,[9] or an appeal to the practice of seventeenth-century poets, or an invocation of the importance of stresses as against numeration of syllables, as in Coleridge's preface to *Christabel*: 'I have only to add that the metre of Christabel is not, properly speaking, irregular, though it may seem so from its being founded on a new principle: namely, that of counting in each line the accents, not the syllables.' There is a certain obscurity in this statement, as we shall see. Nevertheless, poetic practice was including a more liberal use of pauses, monosyllabic feet, and 'real' trisyllabic feet. Meanwhile the philosophy and the

[8] 'Though the time-table has been well known above 150 years, little use has been made of it beyond the pale of music': Richard Roe's preface to *The Principles of Rhythm* (1823), ii. Fussell rightly notes that the classical analogy co-operated with the musical bar analogy to open up the use of more varied feet (*Theory*, 108–9); see also Culler, 'Edward Bysshe', 884–5. Also see Alden, 'The Mental Side of Metrical Form': 'The chief constituent elements of rhythm are generally considered to be stress and time, and neither the stress nor the time of the sounds of verse is indicated by the poet . . . with anything like the clearness which the musician attains through the symbols available for his art' (20).

[9] What Blake and Keats say are typical of late statements in the Romantic reaction. Blake, probably seeing Milton's 'Monotonous Cadence' through the eyes of Bysshe (see Blake, *Notebook* (Oxford, 1973), N88), proclaimed for himself 'a variety in every line, both of cadences and number of syllables' (preface to *Jerusalem*). Keats denounced 18th-century prosody: 'They sway'd about upon a rocking horse, / And thought it Pegasus' ('Sleep and Poetry', 186–7).

images were becoming more appropriate. The notion of organic law seemed able somehow to reconcile the notions of variety and uniformity. But the early Romantic metrists were not much interested in this contrariety because 'law' for them had such bad connotations, those of Newtonian clockwork. Samuel Say, for example, a precursor of the Romantics, celebrated variety and its expressive possibilities, but he took uniformity for granted. Despite his exclamations about 'variety in uniformity' he should not be considered a significant precursor of matured theories of metrical tension.[10] Indeed the Romantic reaction in favour of variety committed another form of the fallacy of misplaced concreteness, by identifying metrical rhythm with the free variations of the spoken language. For the pre-Romantics and Romantics, 'harmony' no longer equals uniformity but comes to equal variety, as in Say and Thomas Sheridan (227). Only in the later Victorian period will the notion finally develop that harmony equals the interplay between variety and uniformity or, in Coleridge's terms, the reconciliation of 'these two conflicting principles of the FREE LIFE, and of the confining FORM' ('On the Principles of Genial Criticism', Essay Third), or elsewhere 'the balance or reconciliation of . . . sameness, with differences; of the general, with the concrete', and so on (*Biographia Literaria*, XIV). Such ideas co-operating with those of Hegel come to fruition in the best pages of Victorian prosody. This dialectical understanding of organic law is the basis of the 'new prosody', which saw a dialectical relation between the abstract metrical form and the individual speech rhythms.[11]

Gradually, therefore, like a pharos light seen through a shifting fog, the nineteenth-century notion of metrical form

[10] Though Fussell celebrates Samuel Say's emphasis on accent, Omond says more perceptively: 'While thus sound about variations, he is less clear as to what they vary from, and ignores fundamental uniformity' (*English Metrists*, 50).

[11] Coleridge implies that one set of dissimilitudes is metre and speech rhythm, but he does not develop the idea. Coleridge and Wordsworth were more interested in defining the function of metre (see below, p. 104); but their insight into its role as an artificial frame, intersecting with meaning in complex ways, was an important contribution to the growth of the prosodist's sense of the metrical abstraction. For a suggestive paragraph by Coleridge on 'the interpenetration, as it were, of metre and rhythm', see his *Miscellaneous Criticism*, 337–8.

begins to come clear. In 1812, John Thelwall, acknowledging the work of the musical scansionists (Steele, Odell, and Roe), notes in passing the 'abstract' or 'skeleton rhythmus, which recognizes only the mere inherent qualities of the elements and syllables arranged'. He distinguishes the skeleton rhythmus from the 'rhetorical rhythmus,' 'that vital and more authentic rhythmus which results from the mingled considerations of sentiment, pause, and emphasis' (x). The image of a 'skeleton rhythmus' will be an important one later. In his 1815 preface to his *Poems*, in a passage missed by the metrical historians, Wordsworth wrote:

Poems, however humble in their kind, if they be good in that kind, cannot read themselves; the law of long syllable and short must not be so inflexible,—the letter of metre must not be so impassive to the spirit of versification,—as to deprive the Reader of all voluntary power to modulate, in subordination to the sense, the music of the poem;—in the same manner as his mind is left at liberty, and even summoned, to act upon its thoughts and images.

This comment may have a certain influence on the notion of 'ordered liberty' which the Victorians will explore. In 1823 Roe, drawing some conclusions from his use of the musical bar, pondered the relation of 'law' and 'license' and wrote: 'In all these cases, the law having been well established, and its influence sufficiently felt, we suppose its continued operation; which prevents the license, if not too long continued, from producing confusion' (9). But if Wordsworth was impressionistic, Roe still sought exact musical measurement.

In 1838, Edwin Guest published his *History of English Rhythms*, which Omond calls 'a landmark of great importance'. It is the first major history of English metres and tries to impose an Anglo-Saxon model on all English rhythms by dividing the verse line into 'sections' or 'versicles' (that is, distinctive groups of accented and unaccented syllables). This system, though it failed to explain accentual-syllabic verse, enabled Guest to list, in an unusually comprehensive way, the very many variations in our standard rhythm, a task which an earlier prosodist like Bysshe would have looked upon as listing sins. Guest also provoked continued reflection on the question: how does the Chaucerian tradition differ from the Old English tradition? How does the metrist avoid the Scylla of accentualism, counting only the stresses, and the Charybdis of eighteenth-century mechanism, counting only the syllables (or, archaically, the quantities)? In 1844, Leigh Hunt devised a better explanation of Coleridge's *Christabel*: Coleridge restored the octosyllabic line to the 'beautiful freedom of which it was capable, by calling to mind the liberties allowed its old musical professors the minstrels, and dividing it by *time* instead of syllables;—by the *beat of four* into which you might get as many syllables as you could, instead of allotting eight syllables to the poor time, whatever it might have to say'. . . . 'The principle of Variety in Uniformity is here worked out in a style "beyond the reach of art".' Poe complained in 1848 that a system of scansion by stresses as proposed by Coleridge for *Christabel* was no system at all. Poe demanded a 'rationale of verse' but used older quantitative notions of 'equivalence' for his own definition.[12]

Nevertheless Poe was right about Coleridge. The lines in *Christabel*, as the Victorians pointed out, varied not only in the number of syllables per line but in the number of stresses per line. For the prosodists who followed Coleridge, the question remained: if uniformity was not to be found in the number of syllables, nor in the number of stresses, nor in the prescribed sequence of classical feet, nor in musical timing, nor in quantitative timing, where was it to be found?

Patmore and the New Prosody

The scene therefore was set for Coventry Patmore, whom

---

[12] Omond, *English Metrists*, 148; Hunt, 55, 58. Gradually we see some progress in understanding the status of time in English verse: from the classicist view that verse time consists in units determined by syllabic quantities, to the 'Englished' classical view that stresses make quantities, to the musical theorists's view that time consists in isochronous bars signalled by the stresses, to the Victorian solution of mental isochrony or abstract spacing. When Coleridge says 'Read him [Massinger] aright, and measure by time, not syllables' (*Miscellaneous Criticism*, 77), he, like Poe, seems to draw on some combination of stress-determined quantities and musical isochrony. On Coleridge, see Whalley and Patterson, and Brogan in *Three Models for English Verse*.

Omond describes as 'inaugurating what I have called the "new prosody" '. In an 1850 review of *In Memoriam*, Patmore had said: 'The want of attention in all recent criticism to the question of metre has resulted from an almost total ignorance of the depth and worth of the intellectual laws on which it depends for its being and effect' (533). Five years later, in a review of *Maud*, he said: 'The interest and attention which the subject of Metre is at present exciting is a curious fact in our poetical history, which seems, in this regard, to have traversed a great cycle of change, and to have returned to an experimental era much resembling that which commenced with Surrey' (517). Patmore sensed that something was up in metrical discussion, backed by the poetic achievements of the golden 1850s. In 1857 Patmore published his major essay on the subject of metre. In it he noted that earlier metrists had 'formed too light an estimate of their subject, whereby they have been prevented from sounding deep enough for the discovery of the philosophical grounds and primary laws of metrical expression' (*Essay*, 6). The essay was first published as a review essay, 'English Metrical Critics', in the *North British Review*, August 1857. Then it was reissued in various modified forms: as a *Prefatory Study* in 1878 to his poem *Amelia* where Ruskin read it and where Hopkins probably read it, perhaps for the second time; then included in his *Poems* of 1879; then made into an appendix entitled 'Essay on English Metrical Law' for the second volume of his *Poems* in 1886, and included in subsequent editions of the *Poems* up to 1906. Hardy would take notes about 1907 from the *Amelia* version of the essay. The essay has been beautifully edited and annotated by Sister Mary Roth.[13]

[13] 'The New Prosody' is Omond's label for the metrical era 1850–1900 (*English Metrists*, Ch. 5, see p. 171); Karl Shapiro follows Omond's description of the importance of the Patmore–Saintsbury line in a good summary article, 'English Prosody and Modern Poetry'. On the history of the editions of Patmore's essay, see Roth's preface to the *Essay*, ix–xii. Punctuation shows that Hardy's excerpts come from the 1878 or later versions of the essay (compare *Literary Notebooks*, ii. 191–2, to Patmore, 17–18, 21–6), and probably from the 1878 or 1879 versions (compare *Literary Notebooks*, ii. 192, to Patmore, 23, 26). Björk assumes that Hardy is copying from the *Amelia* edition, but misdates it as 1872 (Hardy, *Literary Notebooks*, ii. 533).

Following Patmore's essay there occurred a series of discussions and essays, some technical, some popular, which together built up the new notion of prosody. The 1860s culminated the greatest revival of imitation and discussion of classical metres since the late sixteenth century (see below, Chapter 2); and this phenomenon encouraged renewed consideration of the old distinction between *rhythmus* and *ratio*. Discussions continued among several figures associated with the Philological Society in the 1860s and 1870s: Alexander Ellis, Henry Sweet, Walter Skeat, F. J. Furnivall, J. B. Mayor, Thomas Barham, T. Hewitt Key, C. B. Cayley, and others.[14] The number of articles and books published on metre doubled in the 1860s (and again in the 1880s).[15] Skeat would reissue Guest's *A History of English Rhythms* in 1882. Patmore's article was read by Hopkins, who had begun writing about prosody in the 1860s. It was also read by R. W. Dixon and Bridges, who corresponded with each other and with Hopkins and Patmore. Out of this came Hopkins's letters on prosody in the 1870s and 1880s. Hopkins's direct metrical influence at this time was of course limited to his correspondents, but his insights into the new prosody were so incisive that this influence was a significant one. Another contribution was Bridges's various essays and revisions of essays on Milton's prosody, beginning in 1887 in Beechings's edition of Milton's poems with Bridges's appendix on the blank verse of *Paradise Lost*, followed in 1889 by Bridges's supplemental essay on the prosody of *Paradise Regained* and *Samson Agonistes*. In 1893 these essays were made into a book, which was revised in 1901, and again in 1921.

Other important contributions of these years to the 'new prosody' were J. A. Symonds's suggestive discussions of blank verse in 1867 and 1874—which last Hopkins called to Bridges's attention (3 April 1877, *Letters*, 38); David Masson's essay on

---

[14] See, for example, the discussion published in the *Transactions of the Philological Society* (1873–4), 644–5.

[15] The number of essays and books concerned with metre in the nineteenth century are 13 (1800s), 12 (1810s), 16 (1820s), 11 (1830s), 21 (1840s), 21 (1850s), 41 (1860s), 52 (1870s), 120 (1880s), 178 (1890s)—according to T. V. F. Brogan's unpublished 'Chronological Index to English Versification 1570–1980'.

Milton's versification in the first volume of his 1874 edition of Milton's *Poetical Works*; R. L. Stevenson's vivid 1885 statement in 'On Style in Literature: Its Technical Elements'; in America, Francis Gummere's various articles and brief but influential discussion of 'hovering accent' and 'wrenched accent' in his 1885 *Handbook of Poetics*; J. B. Mayor's *Chapters on English Metre* (1886). The most important American contribution was Sidney Lanier's 1880 *The Science of English Verse*. Lanier explained clearly but with too much musical precision how the syllable units (the 'primary rhythm') were grouped by accents into a 'secondary rhythm' organized into music-like bars: 'the poet, after having clearly indicated . . . this normal time-value of each bar, may then go on to vary the individual time-values of the constituent sounds in any given bar at pleasure' (111). According to Omond, 'the "new prosody" takes in his book a step which can never be retraced'—presumably by so clearly stating the musical bar analogy (*English Metrists*, 202). In England in 1880 Ruskin, probably inspired by his recent reading of Patmore's essay, helped revive the musical bar analogy in *Elements of English Prosody*; and either Lanier or Patmore probably convinced Hopkins to use the musical bar analogy in his 1883 'Author's Preface' (*Poems*, 45).

In 1887 Saintsbury's *History of Elizabethan Literature* appeared and was followed by other writings which eventually led to his *History of English Prosody* in three volumes (1906, 1908, 1910). T. S. Omond published his first article on prosody in 1875; in 1897, after a long silence, he began an extensive series of essays on prosody, followed in 1903 both by his *Study of Metre* and also his *English Metrists* which was revised in 1907, and again in 1921.[16]

The late Victorian period also witnessed a great body of scholarship focusing 'on the elucidation of the versification of England's three greatest poets, Chaucer, Shakespeare, and Milton' (Brogan, *Three Models*, 89). The scholars involved

---

[16] For Omond's other works, see Brogan, *English Versification*, 38, 183–4, 201–3, 348, 378, to which should be added 'English Prosody', *Gentleman's Magazine*, 284 (Feb. 1898), 128–44, and 'The Limits of Verse-Length', *Living Age*, 258 (11 July 1908), 119–22.

debated the issue of real versus elided trisyllabic feet, and thus
rejuvenated a discussion that had been current since the late
eighteenth century (*Three Models*, Chapter 2, 54). (Hardy's
metrical revisions, eliminating contractions, show he eventually sided with the trisyllabists.) Indeed the debate between the
trisyllabists and the strict syllabists (like Bridges) was like the
opposition between Gildon with his feet and Bysshe with his
syllable-counting, at the beginning of the eighteenth century.
Many of the figures I have cited above participated in this
scholarship; but what interests me here are those persons,
scholarly and lay, who contributed to a growing consensus
about the abstract nature of metrical form.

The above works, from Patmore to Omond, seem to be the
key landmarks of the new prosody—to which might be added
Raymond Alden's *English Verse* (1903). There were many other
contributions, of course, one of which I would emphasize,
partly because it was read by Hardy, partly because it
summarized the insights of Patmore and the others in a
definitive way: 'English Prosody', an unsigned review article of
Bridges, Omond, Saintsbury, and William Stone, published in
the *Quarterly Review* in July 1911. The author of the article was,
in fact, Basil De Selincourt (letter to me from John Murray, the
publisher, 24 February 1984).

In his notebook, Hardy summarized the insights from
Patmore's landmark article:

English metre.—. . . 'The function of marking, *by whatever means*, certain
isochronous intervals. . . . Two indispensable conditions of metre. 1:
that the sequence of vocal utterance . . . shall be divided into equal or
proport[ionate] spaces; 2: that the fact of that division shall be made
manifest by an 'ictus' or 'beat.' (*Literary Notebooks* ii. 191.)

(On Patmore's qualification of 'isochronous', see below.)
Hardy did not copy the rest of the sentence which continues:

. . . 'ictus' or 'beat', actual or mental, which, like a post in a chain
railing, shall mark the end of one space, and the commencement of
another. . . . Yet, all-important as this time-beater is, I think it
demonstrable that, for the most part, *it has no material and external*

existence at all, but has its place in the mind, which craves measure in everything. (*Essay*, 15.)

Hardy will also copy from Basil De Selincourt's 1911 article:

The principle which we deduce may be expressed thus: so long as the structure of a verse shows either in itself or in its context the number of accents which it ought to have and the places where they ought to fall, so long as the mind hears the implied accents in their places, the number and position of the accents which naturally occur is of no consequence.[17]

The idea of these abstract spaces is the key insight of the new prosody. Interestingly, both visual and aural analogies were invoked for these recurrent mental accents. The common solution was that the metrical accents were 'mental' or 'felt' rather than 'heard'. Raymond Alden, an important disciple of Omond, will entitle a 1914 essay 'The Mental Side of Metrical Form'. What Patmore called intervals, Omond called time-spaces, and Saintsbury called feet—all ways of naming these mental spaces.[18]

The sense of this abstraction is what separates David Masson in 1874 from Robert Latham in 1841. Masson adopts Latham's system of notation. However Latham's notion of 'groups' of syllables marked by accent remains arithmetical. Masson's notion of equivalence approaches Saintsbury's: *'what the cultured ear would accept as equivalent'* (Masson's italics, cxiv). In 1901, Thomas Goodell will apply this sense of an abstraction to classical quantitative verse!

[17] De Selincourt, 93. 'No consequence', as we shall see, is too strong a claim. Hardy's notes from Patmore and De Selincourt are in *Literary Notebooks*, ii. 190–2, 209. The pattern of copying in *Literary Notebooks* suggests that Hardy took notes from Patmore about 1907 and from De Selincourt in 1911. About Patmore's essay, Hardy wrote in his notebook: 'some of these ideas are suggestive—others doubtful' (*Literary Notebooks*, ii. 193). This evaluation very much parallels Omond's (*English Metrists*, 175). De Selincourt's passage which Hardy copied continues: 'And we may at once add that, in regard to the number of syllables in the foot, a similar principle is to be applied: the verse structure must make it clear how many syllables are to be expected, how many, that is, are to be taken as normal; the poet may then substitute more or fewer at his discretion'.

[18] Also see Patmore, *Essay*, 15, and Roth's note 15.30 of the *Essay*; Omond, *English Metrists*, 172.

The particular rhythm is conceived as a mould or pattern to which a pliable material is made to conform; the pattern exists in the mind of the ῥυθμοποιός, and receives objective audible existence only by embodiment in a ῥυθμιζόμενον.[19]

Hopkins's contribution to this discussion is significant—and necessary before he could develop his notion of sprung rhythm. In his 1873–4 lecture notes, he said that in Milton's 'accentual counterpoint' in *Paradise Regained* 'the beat of the line has to be carried in the mind: it is not expressed.' In a letter to Bridges on 5 September 1880, he scans a Wyatt line and imposes his symbols:

His rhythm . . . is very French and lightsome, lighter than Surrey's and weaker, and that appears here; for instance I think he wd. scan— 'it sitteth me near', that is, really, 'it sitteth me near' or, as I like to write it, 'it sitteth me near', the black ball marking the real or heard stress, the white the dumb or conventional one.

In a letter to Dixon of 14 January 1881, he describes 'that conventionally fixed form which you can mentally supply at the time when you are actually reading another one'.[20]

Hopkins's later speculations about abstract form were clearly influenced by Patmore. He may have been familiar with Patmore's ideas at an early date.[21] In a letter to Patmore of 7

---

[19] Goodell, 102. This abstraction, Goodell claimed, was more powerful than any measurement of quantities. Syllabic quantities in fact had 'considerable elasticity'; thus 'precise measurements : . . are illusory' (100, 181). Latham also used the term 'metrical fiction' in a way that may have influenced Bridges, but Latham is not cognizant of the full abstraction of that fiction: 'A metrical fiction, that conveniently illustrates their structure [i.e. the trisyllabic verse of *Christabel*], is the doctrine that they are *lines formed upon measure x a x, for which either x x a or a x x may be substituted, and from which either an a x or x a may be formed by ejection of either the first or last unaccented syllable*' (Latham, 489). An additional interesting item of Victorian prosody was first printed in 1980—George Eliot's 1869 essay 'Versification', which was apparently influenced by her reading of Symonds's 1867 essay 'Blank Verse' (Eliot, 286). Eliot notes the abstract nature even of musical rhythm: 'a large proportion of these beats are perceived by the inward sense only, and are not represented in sounds that strike the tympanum' (287).

[20] Hopkins, *Journals*, 282; *Letters*, 109; *Correspondence*, 41. Never mind that Hopkins is using a faulty edition of the line: see Norman MacKenzie, 'Hopkins and the Prosody of Sir Thomas Wyatt'.

[21] See Edward Stephenson, 'Hopkins' "Sprung Rhythm" and the Rhythm of *Beowulf*', p. 108 n. 20.

November 1883, Hopkins discussed Patmore's essay with some
reservations. He distinguishes more clearly English accent and
Greek tonic accent ('essentially it was pitch'), which Patmore
had discussed in the passage quoted above by Hardy. Hopkins
does not dispute that passage's discussion of the function of
'ictus' which has no 'material . . . existence at all' in creating
'equal or proportionate spaces'. But Hopkins makes a valuable
addition to Patmore's argument: 'The stress, the *ictus*, of our
verse is founded on and in the beginning the very same as the
stress which is our accent . . . and when discrepancies do arise
they begin so naturally that people may well not notice them.'
That is, people may not notice, until 'later', the discrepancy
between the metrical ictus and the speech stress. Thus in a line
like

> a penniless adventurer is often in extremities

'perhaps people wd. not notice that every other strong beat,
every fourth syllable, that is, is really scarcely marked at all, so
inevitably does the mind supply it.' 'From the same kind of
sentences', Hopkins continues, 'may also arise the *blank stresses*,
as I am accustomed to call them, of the ten-syllable line and
other lines; for in fact in Milton few lines have five real stresses,
one or two being blank, though in idea there are always five'
(*Further Letters*, 328). Patmore said very well when he replied to
Hopkins: 'much of the substance of your very valuable notes
will come in rather as a development than as a correction of the
ideas I have endeavoured—with too much brevity perhaps—to
express' (*Further Letters*, 334).

To discover the abstract dimension of metre was not
necessarily to define it well. Patmore himself in his essay
sometimes lost the sense of it and had to be corrected by
Omond for confusing the mental ictus with the real ictus, or the
metrical accent with the speech stress.[22] Omond in turn fell in

---

[22] Generally in this book I have used the word 'stress' to denote the natural speech
stress, 'accent' to denote the metrical accent. But such a distinction of words is not
maintained in traditional metrical discussion; and in my own discussion, the difficulty
of distinguishing the words shows the interdependent nature of the two notions in
actual verse instances. Lexical stress becomes part of a second-order level emphatic

love with his favoured way of describing the rhythm as organized into temporal intervals or 'time-spaces', as though they could be measured by a clock (*Study*, 3–4). Masson's uncertainty about the distinction makes him a transitional figure from the old to the new prosody, as when he is puzzled by the now 'common notion of accent, which makes it a mystic something, distinct from stress, strength, or anything that can be perceived in actual enunciation' (cxx).

But Omond at his best thought of these 'time-spaces' as abstract. They did not depend on the steady signals of stess: 'those who make accent the constitutive principle of English metre seem to confound this *ictus* with the structure it illustrates, the period with the bell that calls attention to it'. ('The absolute length of periods matters very little.') He quotes Patmore on the 'mental beat' having 'no material and external existence at all' and agrees that the 'time-beater of rhythm is mental'. Saintsbury acknowledged that his 'feet' were practically identical with Patmore's intervals. Indeed in his 1850 review article on *In Memoriam*, Patmore had used 'feet' in a way that anticipated Saintsbury. In his copy of *The Rhymer's Lexicon*, Hardy read Saintsbury's introduction: ' "Principle of equivalence" has governed our verse. For "feet" one might use "isochronous interval" but this is clumsy; "Space" by itself might do; but it is really unnecessary.'[23]

Theorizing about these spaces and mental beats is a tricky business, as the abstraction gets confused with the actual elements of a given verse line. It is easy to find one's opponent materializing in the fact. One must read these essays carefully. Early in the 1857 essay, Patmore congratulates Steele for defining 'the true view of metre, as being primarily based upon isochronous division by ictuses or accents'; but later in the essay Patmore will qualify: 'the equality of proportion of metrical

---

stress, and both in turn become part of a third-order level of metrical accent. The lexical and emphatic stresses give rise to metrical accents, and can also be influenced by metrical accents. See below on this hermeneutical circle.

[23] Omond, *Study*, 24, 29; *English Metrists*, 172 (also see 155 on confounding 'verbal accent with metrical stress'); Saintsbury, iii. 439–40; Patmore, '*In Memoriam*', 534; Saintsbury, *Rhymer's Lexicon*, xviii.

intervals between accent and accent is no more than general and approximate.' De Selincourt insisted on this qualification of the principle of 'isochronous intervals': 'It does not matter what is the length of the intervals, provided that they seem the same.' In 1918–19 there would occur an interesting series of letters on metre in the *Times Literary Supplement*, several of which Hardy cut out and put in the back of his copy of Guest's *History*. The letters debate the issue of exact versus approximate, material versus abstract equivalence—until D. S. MacColl concedes on 6 February that the metrical beats are 'in theory . . . equidistant; in practice not always'.[24]

Patmore thus applied a new philosophical sense to the old metrical issues. He had taken the *disjecta membra* of previous theorists—Puttenham, Mitford, Steele, Guest, and others—and infused them with a sense of what he called 'intellectual laws' and 'philosophical grounds' ('*In Memoriam*', 533, *Essay*, 6). The philosophy which he cited most often was the dialectic of life and law in Hegel. Indeed Hegel was not only a major source of the general notion of dialectic, but also the first nineteenth-century source of the idea of metrical counterpoint. Hegel had sketched the notion as early as *The Phenomenology of Mind* (1807, published in full in a French translation in 1867–9). Here Hegel added a dialectical element to traditional speculation about classical metres and the relation between the pattern of quantities and the speech stresses. Citing the 'conflict . . . between metre and accent in the case of rhythm', Hegel noted: 'Rhythm is the result of what hovers between and unites both' (120). Only an analogy here, the notion is fully developed in Hegel's *Aesthetics*, published in a French translation by Bénard in 1840–52. A partial English translation began in 1867 in the *Journal of Speculative Philosophy*. Patmore would read them in Bénard's version, before writing his *Essay*.

[24] Patmore, *Essay*, 5, 21; De Selincourt, 77; see Hegel, *Aesthetics*, 1017: 'poetry . . . need not be subject, so abstractly as is the case with a musical beat, to an absolutely fixed measure of time.' Also see Lascelles Abercrombie, *Principles of English Prosody*: 'The introduction of duration into accentual-prosody seems to be due to a desire to capture in definable terms the somewhat elusive fact of equivalence, which is the essential fact in the existence of accentual rhythm as metre' (137). On the *TLS* letters, see below, n. 37.

Hegel discusses music before he discusses poetry. In music, 'the counter-thrust between the rhythm of the bar and that of the melody comes out at its sharpest in what are called syncopations' (Knox translation, ii. 918). 'The bar, rhythm, and harmony are, taken by themselves, only abstractions which . . . acquire a genuinely musical existence only through and within the melody' (ii. 931). This dialectic between the abstract bar-like rhythm and the melody has a certain similarity, Hegel suggests, to what we find in poetry, between the 'feet' and the phrases (ii. 918). About classical verse Hegel then says: 'we have on the one side the metrical accent and rhythm, and on the other side ordinary accentuation. Both of these means are intertwined with one another, without mutual disturbance or suppression, to provide a double variety in the whole' (ii. 1026–7). Turning to the German accentual line, however, Hegel did not see any precise abstract structure there; he concluded that with the loss of measured quantities, the art of intertwining was lost (ii. 1019–21). Here is where the English Victorians would step in and find a parallel art, between the pattern set up by the accentual-syllabic metre and the normal speech rhythm.

Translations of Hegel into English coincided with the rise of the new prosody: the *Philosophy of Spirit*, translated in part in 1868; the *Science of Logic*, translated by Sloman in 1855 (French translation by Sloman in 1854); the *Encyclopaedia of the Philosophical Sciences* translated in part in 1869 (French translation by Vera in 1859); the *Philosophy of Right*, translated by Sanders in 1855; the *Lectures on the Philosophy of History*, translated by Sibree in 1852. What Patmore added to Hegel most significantly was the notion of abstract spacing produced in the modern accentual languages by a pattern of stresses and non-stresses. Patmore frequently cited those sections from Hegel which I have quoted above.[25]

---

[25] On Hegel, see Kurt Steinhauer, *Hegel: Bibliography* (Munich, 1980). On Patmore and Hegel, see Roth's note 7.3 and *passim*. Hegel also described well the relation of Greek and Latin, Greek quantities making the early Latin accentual verse adapt itself to a quantitative measure which the Romans then managed more strictly than the Greeks (*Aesthetics*, 1024); then Hegel described the emergence of accentual rhythm and

Given this notion of a metrical abstraction, the poet can then take liberties, up to a point (to be discussed below), with all the traditional elements of metre—syllables, stresses, temporal isochronies, feet. Thus Hardy copied from Bridges: 'The main effectual difference between the rhythms of the old metrical verses & of fine prose is that in the verse you have a greater *expectancy* of the rhythm . . . & the poet's art was to vary the expected rhythm as much as he could without disagreeably baulking the expectation' (Hardy's italics). And from De Selincourt, Hardy copied: 'The art of the poet consists in introducing variation upon this basis of equality.'[26]

It may be the virtue[27] of the new prosody that it did not clearly specify the point at which an actual violation of the metre would occur. Here the old orthodoxy was better equipped to answer—at the price of over-simplification. Bridges argued somewhat clumsily that in blank verse those variations were allowable where 'the interruption is not long enough, and the majority of verses sustain the impression of the typical form'. Saintsbury said that the norm 'must not be confused by too frequent substitutions'. The restraints he did suggest—only one strong stress in a trisyllabic foot, monosyllabic feet reserved for the first and last feet unless next to a strong pause, no more than three unaccented syllables in a row in combined feet—represent a lingering conservatism which Hopkins would challenge. Masson also proclaimed: 'the acceptability of a line to the ear . . . is by no means in the inverse proportion of the number of its variations from the normal' (cxix). And Saintsbury conceded that his rules 'are not imperative or compulsory precepts, but observed inductions from the practice of English poets. He that can break them with success, let him.'[28]

---

rhyme in medieval Latin verse (1025). Wellek in his *History of Modern Criticism: 1750–1950*, vol. ii (New Haven, 1955) traces the German Romantic genealogy of Hegel's ideas; but Hegel's notion of 'the clash between the metrical pattern and the rhythm of prose'—cited briefly by Wellek on p. 323—seems unique to Hegel.

[26] Hardy's *Memoranda Book II*, in *Personal Notebooks*, 62–3, from Bridges's 1922 'Humdrum & Harum-Scarum'; De Selincourt, 79, Hardy, *Literary Notebooks*, ii. 209.

[27] 'Or vice', Brogan comments, in notes to me on this manuscript, 13 June 1984.

[28] Bridges, *Milton's Prosody* (1921), 42, substantially same phrasing in the 1893 and

Symonds said of the iambic pentameter: 'so various is its structure that it is by no means easy to define the minimum of metrical form below which a Blank Verse ceases to be a recognisable line'. Masson noted that English blank verse admits trochees, spondees, pyrrhics, and trisyllabic feet 'in almost any place in the line' (cxviii, cxxii); 'the number of accents, unless in a peculiar sense of accent, not realized in actual pronunciation, is also variable' (cxx). Stevenson exclaimed: 'in declaiming a so-called iambic verse, it may so happen that we never utter one iambic foot'. Bridges concluded that in Milton's *Paradise Lost* 'there is no one place in the verse where an accent is indispensable'. This led Bridges to define Milton's prosody there as syllabic—that is, so many syllables per line, with extra syllables reduced through the 'fiction' of elision.[29]

But generally the Victorian prosodists believed metre governed both accents and syllables, and relied largely on intuition, based on a knowledge of traditional poetic lines, of what

1901 editions; Saintsbury, i. 82–4; Saintsbury, *Historical Manual*, 30; in the *History of English Prose Rhythm* (462) Saintsbury also says: 'the substitution of dactyl for iamb, and anapaest for trochee, with the consequent juxtaposition of the two in each case, always, or almost always, leads to jangling and jarring' in verse. Mayor, overly severe against Symonds's 'intuitive' notion of equivalence, backs himself into too conservative a corner, denying more than 3 syllables to a foot (53). More reasonably perhaps, he suggests that 'the limit of trochaic substitution was three out of five' in the iambic pentameter (72). Jespersen would later suggest a mathematical way of determining the degree of variation from the norm. An initial trochee in an iambic pentameter line meant that 'the ear is really disappointed at one only out of ten places', i.e. the fall between the 2nd and 3rd syllables where one expected a rise. And degree of disappointment must be considered. 'A beginning 4114 is comparatively poor, but 4314 or 4214 does not sound badly', where 4 is the strongest degree of stress, 1 the weakest degree (78). Jespersen does not give the number and degree of disappointments that would constitute unmetricality.

[29] Symonds, *Blank Verse*, 12; Stevenson, 555; Bridges, *Milton's Prosody*, 39; Bridges, 'Letter', 23. Bridges's theory swayed later Milton scholars (see Brogan, *Three Models*) because he showed that what looked like 'real' trisyllabic feet in *Paradise Lost* had in fact 2 elidable syllables in the unaccented position. 'Blank verse which admits into such places *any* kind of unaccented syllables, whether elidable or not, ceases to be syllabic verse and becomes so far accentual' (my italics) (*Milton's Prosody*, 17). By 'fiction' Bridges meant that 'Milton came to scan his verses in one way, and to read them in another' (35), i.e. he pronounced all his syllables but still maintained the possibility of elision. Bridges thus distinguished Milton's 'prosody' with its fictional rules and Milton's rhythms. For *Samson Agonistes*, however, written with 'true trisyllabic rhythms' (53–4), Bridges went with the standard Victorian theory.

variations were possible. Exactly how the metrical law seemed to permit some variations and disallow others—or more correctly, since a law cannot be varied, how some speech rhythms observe the law and others violate it—was a question re-posed with influential clarity by Halle and Keyser in 1966. It is a question that has yet to be fully answered. But the Victorians did suggest one way in which the metre clearly breaks. In 1889 W. H. Browne will suggest: 'Any variation is allowable that does not obscure or equivocate the genus; but any that suggests another genus is not allowable' (198). Stevenson had adumbrated this point four years earlier (555). The Victorians suggested that one metre holds up to a certain point and then breaks and another takes over. This notion represents a giant advance over Bysshe, who had found the line 'Apart let me view then each Heavenly Fair' 'burlesque' but did not know why (7). Like Halle and Keyser's crux line, 'Ode to the West Wind by Percy Bysshe Shelley', it is a questionable iambic pentameter because it tends to fall into a different rhythmic *Gestalt*: the anapaestic tetrameter.[30]

The question remains for us: when does a line exceed the metrical expectation so much that it seems to stumble awkwardly? Can this violation be distinguished from an interesting variation? Can a metrical line be judged as an Olympic judge might judge the grace, and mistakes, and extraordinary figures, in the timing of a dancer. Incidentally, a popular late Victorian image of grace, and of the dialectic reconciliation of form and matter in the Coleridge tradition, was the dance. The last stanza of Yeats's 'Among School Children' is a kind of summary of this tradition: ideally, how can we know the dancer from the dance? But do some metrical dances fail? And why? When does the dancer fall down?[31]

[30] Halle and Keyser, *English Stress*, 139. Later prosodists in the twentieth century tended to rely on statistical frequency of occurrence, but statistical frequency is not the same as law though it may give rise to the sense of law. Tarlinskaja determines what she sees as 'quantitative indices' (14), 'metrical thresholds' at which 'a poem passes over to another system of versification or another (adjacent) meter' (8).

[31] Roth cites the dance analogy in Patmore, Mitford, and others (Patmore's *Essay*, 62 n. 10.11). To Roth's list should be added Yeats and Pope: 'True ease in writing comes from art, not chance, / As those move easiest who have learn'd to dance'.

The Victorians did not answer these questions, though they suggested some directions. They also provided us with the concepts which stand behind our best textbook accounts of metre now. It is to the Victorians, more than to Jespersen (who was immersed in the English Victorian scene) or Jakobson and the formalists or Richards or Lascelles Abercrombie that we first owe the statement of that principle which the Victorians called the 'double structure' of the metre and the language, or what Bridges called the skeleton and the flesh (see below, Chapter 2). The Victorians were the ones who first spoke of interplay and counterpoint and syncopation and modulation in the modern sense. In his 1885 essay, Stevenson said that the true versifiers give us 'a rare and special pleasure, by the art, comparable to that of counterpoint, with which they follow at the same time, and now contrast, and now combine, the double pattern of the texture and the verse'.[32]

Thus from Patmore Hardy copied the comment that 'there seems to be a perpetual conflict between the law of verse and the freedom of the language, and each is incessantly, though insignificantly, violated for the purpose of giving effect to the other'. Saintsbury objected to the notion of 'conflict' and insisted on co-operation, 'ordered liberty'. Bridges described the beauty which accentual speech rhythms gain 'from interplay with a fundamental metrical form' (by which he meant a classical quantitative form). He also noted about *Samson*

---

[32] Stevenson, 552. The preferred modern term for the relation between metrical scheme and speech rhythm seems to be 'syncopation': see, for example, John Hollander, *Vision and Resonance*, 21–2. Bridges, influenced by his early correspondence with Hopkins, early claimed that the iambic structure of *Samson Agonistes* is 'maintained as fictitious structure . . . to be imagined as a time-beat on which the free rhythm is, so to speak, syncopated, as a melody,' *Milton's Prosody*, 1893 edn., 35. Hopkins also used the term 'syncopation' in his 7 Nov. 1883 letter to Patmore (*Further Letters*, 327–8). Also see Saintsbury, ii. 245, 248. The first association of poetry with the word 'counterpoint' may be in Daniel Webb's brief criticism of the Greek hexameter, that it is like 'our modern counterpoint' where expression is sacrificed to artifice (*Observations*, 118). But Webb's insight into this parallel is limited, like Say's: see above, n. 10. Jespersen had a long familiarity with the English philological scene. As a schoolboy he read Müller and Whitney. Shortly he was reading Sweet and Ellis, whom he would meet in London. In 1893 he was professor of English at the University of Copenhagen. In his works he would often cite Archibald Sayce and Spencer's ideas on style. See *Linguistica*, 2, 3, 87.

Agonistes: 'where the "iambic" system seems entirely to disappear, it is maintained as a fictitious structure and scansion, not intended to be read, but to be imagined as a time-beat on which the free rhythm is, so to speak, syncopated, as a melody'.[33]

Henry Newbolt, generalizing in a 1912 *English Review* article, said: 'What there has always been—whether in Greek or Latin, or in modern poetry—is an antagonism, a balance, a compromise, between the metrical ictus, the drum-beat which I have imagined, and the common speech-rhythm of the language'; 'most of the beauty of the lines and all their variety is gained by the skill with which the woof of speech-rhythm is continually thrown athwart the warp of the metrical type' (662–3). I mention this article because Hardy was a long-time friend of Newbolt. Newbolt discussed poetry with Hardy, included Hardy in his *English Anthology* and elsewhere, received many personally inscribed Hardy works, sent Hardy his own poems, and participated in various award ceremonies for Hardy. It is likely that Hardy read Newbolt's essay not too long after he read De Selincourt's article.

Hopkins also used the term 'counterpoint' but in a more restricted way than Stevenson. In 'common' rhythm, when 2 or more feet in a row are reversed, a counter rhythm is created which we hear along with the standard rhythm ('Author's Preface'). This is perhaps a more accurate notion of counterpoint but, even so, a somewhat loose analogy from music where we physically hear two melodies at once. Modern textbooks took over the term in Stevenson's sense rather than in Hopkins's to stand for the general interplay of abstract form and speech rhythm.

The new prosodists generally agreed that metre is first recognized by the way it emerges from the natural stress rhythms of the words. The main stresses of substantive words, like nouns, verbs, adjectives, adverbs are the main determinants of metrical rhythm; the unaccented places are filled up

---

[33] Hardy, *Literary Notebooks*, ii. 191, Patmore, *Essay*, 9; Saintsbury, iii. 439, 296 (also see i., 345); Bridges's *Milton's Prosody*, 87, 55 (this last in the 1893 edn. also, 35).

by subsidiary function words like articles, pronouns, auxiliary verbs, prepositions, conjunctions, and so on. In Longmuir's preface to Walker's *Rhyming Dictionary*, Hardy read: 'the words suggestive of persons, animals, things, and actions are the most important; other words do little more than connect these principal words or point out their qualities and relations. . . . This naturally suggests the words in a sentence on which the emphasis would be placed.' A 'regular order' of such 'emphatical words or accented syllables' 'would arrest the attention. . . . Syllables thus resolve themselves into emphatical or strong, and unemphatical or weak; or long and short, as they have otherwise been named' (viii–ix). But the minor words can also influence the rhythm in various subtle ways; and the stresses of the major words can also be affected by the rhythm. The distinction between major and minor category words— metrically speaking—has yet to be clearly made. Indeed absolute distinction may be impossible because of the way in which metre interacts with the language and influences the distribution of accents.

The idea that metre emerges somehow from the natural stress rhythms was consistent with the ancient notion that poetic metre was only a regularization of the metre implicit in prose. Aristotle had said it in the *Poetics*, and the notion was restated with remarkable frequency in the Victorian period. Hardy copied from Patmore: 'as dancing is no more than an increase of the element of measure which already exists in walking, so verse is but an additional degree of that metre which is inherent in prose speaking'. '*Perfect poetry*', Hardy copied, '*and song are, in fact, nothing more than perfect speech upon high and moving subjects.*' Hardy also copied Patmore's controversial claim that the measure of English verse '*is double the measure of ordinary prose*'. Hardy agreed with Herbert Spencer's theory 'on the origin of music', that song rhythms originally arose from emphasizing and intensifying the peculiarities of ordinary speech when used to indicate excited feeling, such peculiarities as 'loudness, timbre or quality, pitch, intervals, and rate of variation'. In Lewes's *The Story of Goethe's Life*, Hardy read: 'Impassioned prose *approaches* poetry in the rhythmic impulse of its movements; as impassioned speech in its various cadences also approaches the intonations of music.' 'Rhythm varies gradually and imperceptibly through numberless gradations . . . till it disappears in the pedestrian progress of average prose', Hardy clipped from an 1899 *Academy* article. Hopkins echoes the tradition: 'Poetry is in fact speech only employed to carry the inscape of speech for the inscape's sake'.[34]

The verse stanza must sometimes be read in its entirety before the metre can be defined: 'the rhythm of the lines must be interpreted by the general rhythm of the piece' (Mayor). 'The metre is determined by *the prevalent foot*, and that cannot always be ascertained till a few lines have been read' (Abbott and Seeley). Early on, Patmore had noted: '*Thus we see that an entire line may be in common or triple cadence, according to the cadence of the context*' (Patmore's italics). Edwin Abbott and J. R. Seeley gave an interesting example:

Thus in Michael Drayton's 'Agincourt' we might read the first three lines, and not perceive the metre till the fourth:—

> Fair stood the wind for France,
> When we our sails advance,
> Nor now to prove our chance,
>   Longer will tarry.

Here it might naturally be supposed, from the first three lines, that . . . dissyllabic metre with three accents was intended; but the fourth line, which is clearly trisyllabic, makes it doubtful whether the first three lines should not be treated as trisyllabic with two accents:—

> Fáir stood the | wínd for France.

[34] Patmore, *Essay*, 10, 17, 26; Hardy, *Literary Notebooks*, ii. 191–2; i. 51, 301 (the quotation is from Guerney's summary of Spencer); Lewes, 234; Hopkins, *Journals*, 289. On the issue of poetry being double the measure of prose, see Shapiro and Beum, 34; for Hardy's reading of Lewes and the *Academy* article, see *Literary Notebooks*, i. 259 ff., ii. 93. Also see Abbott and Seeley, *English Lessons*, 146: 'Now just as the voice rises from (a) conversational non-modulation to (b) rhetorical modulation, and from modulation to (c) singing, so the arrangement of words rises from (a′) conversational non-arrangement to (b′) rhetorical rhythm, and from rhythm to (c′) *metre*.'

The metre seems to be the same as . . .

> Cannon to right of them,
> Cannon to left of them,
> Cannon in front of them,
>   Volleyed and thundered.
>
>   Tennyson[35]

Hardy copied a relevant passage from De Selincourt's article (90): 'As Mr Omond says . . . "The line 'How happy could I be with either', actually varies in metre according as we emphasize the word 'I' or leave it unimportant" '. To emphasize 'I' is to see the line shift, like a *Gestalt* puzzle, from an iambic tetrameter to an anapaestic trimeter.

The Victorians tended to agree that the pattern can influence the normally unstressed words: Rhythm 'can impose itself upon recalcitrant material. Language is full of recalcitrances' (De Selincourt, 76). In his learned study of Chaucer (1884, translated into English in 1901), Ten Brink believed that in some cases the metre caused the normal speech stresses to shift. In other cases, Ten Brink thought that either the normal speech stress made the metre change or, better, that the poet achieved the effect of 'veiled rhythm—level stress' where 'the portion of the verse in question makes the conflict between accentuation and rhythm bearable by the very fact that it preserves the consciousness of the rhythmical scheme'. Ten Brink's distinction was a bit too neat; but the questions of how much stresses 'hover' or are tilted by the metre have never been resolved.[36]

   T. B. Rudmose-Brown asked: 'Why does this line (if it does) strike the ear (or the mind) as a verse, and as a verse of a

---

[35] Mayor, 111; Abbott and Seeley, *English Lessons*, 165; Patmore, *Essay*, 39; Abbot and Seeley, *English Lessons*, 165–6. An early discussion of such convertible metres is Latham's, p. 489.

[36] Ten Brink, 190, 192. The Chatman–Stein dispute on hovering stress was more or less resolved in Chatman's partial concession that a line may be 'capable of two actualizations' (*A Theory of Meter* (The Hague, 1965), 149 n. 38). On 'tilting' see Wimsatt, 'The Rule and the Norm', 209. The 'tension' dispute also seems resolved: see Wimsatt, 'The Rule and the Norm', n. 24, and Halle and Keyser, *English Stress*, 142, 176. Attridge, *Rhythms*, has extensive discussion on 'promotion' and 'demotion', 164 ff.

particular type? Does it do so because we mentally wrest it into our preconceived time-scheme, although we read it differently, or because, with its natural prose rhythm, and as read, it does actually fit our time-scheme?' (Omond and Rudmose-Brown, 'Inverted Feet', 498.) In answer the Victorians were suggesting a sort of hermeneutical circle. The metre must be induced from the given language and imposed on that language. Once perceived, the metre enters into a dialectic relation with the language.

The nature of the relationship between the verse skeletons and the speech rhythms, the manner in which we 'hear' the metrical pattern in the speech rhythms, was much in dispute. The incident of Ruskin and his mother is a kind of *locus classicus* in Victorian discussion of the issue. As a child Ruskin insisted on sing-songing the trimeter hymn line: 'The ashes *of* the urn'. His mother insisted he take the sounded accent off 'of'. Ruskin obeyed, but only later when he read Patmore did he understand why his mother was right. Ruskin reported the incident both in *Fors Clavigera*, in a letter of 1873, and in *Praeterita* (1885–9), chapter 2. In the former version Ruskin said of his mother's success: 'had she done it wrongly, no after-study would ever have enabled me to read so much as a single line of verse'. After Ruskin read Patmore's *Essay* in the 1878 edition, he praised the essay for 'declaring what I now believe to be entirely true (though entirely contrary to my—up to this time—strongly held opinion) that verse must "feel, though not suffer from" "the restraint of metre"'. De Selincourt's comment on the incident seems, again, most perceptive: 'The felt, the implied, accent is here so obvious that young Ruskin could not be prevailed on to let it go; it required the sagacity of a grown-up person to perceive that there is no need to proclaim things that proclaim themselves.' Mayor cited those normally unaccented syllables 'which, if the verse were mechanically regular, would have had a word-accent, and to which therefore the general influence of the rhythm may seem to impart a sort of shadow of the word-accent'. Masson is characteristically honest: 'The matter [of the reality of English spondees] is complicated (as indeed it is in the Pyrrhic) by the delicate question of what the distinction is between accent and mere stress, strength, or quantity.'[37]

The Victorians appreciated that it is important to know that a metrical scheme can put pressure on 'of' and can influence performance in certain cases. Bridges in 1891 clarified the case of *Christabel* by noting in it the pressure of the abstract scheme: Coleridge 'merely *states* that there are 4 stresses in every line. There are not always 4. The reader has to pretend there are, e.g. "Is fástened tó an ángel's féet" (*Selected Letters*, i. 215, a point developed in *Milton's Prosody*, 1893). We distribute the metrical accents in proportional ways, and the distribution may put some pressure on some of the normal speech stresses. This is the genius of our metrical rhythm and makes it capable of such great and subtle expressiveness. A conversation Hardy had with his printer illustrates the point. Queried as to why he used the spelling 'Valency' in 'A Death-Day Recalled' and 'Vallency' in 'A Dream or No', Hardy replied: 'The two spellings are because of the two pronunciations, which the metre requires. The river is called either way indifferently' (*Complete Poetical Works*, 491). The differences in the two poems, respectively, are:

Thín Valéncy's river

and

Or a Vállency Válley

(In the 1930 version of 'A Death-Day Recalled', however, the printer had his way, for we find 'Thin Vallency's river'.) In his revisions, we can also see Hardy pushing around the accents of

[37] Ruskin, *Works*, xxvii. 617, xxxvii. 253; De Selincourt, 93; Mayor, 36; Masson, p. cxviii. In the light of these themes, it is exciting to follow the series of letters occurring in *TLS* in 1919, if only because they represent a later matured discussion of issues which Patmore had raised over sixty years before. Hardy cut out the clippings of some of these letters and put them in the back of his copy of Guest (Holmes, Listing). On 9 Jan., Sturge Moore wrote to *The Times* complaining that the metrical scheme 'tells us less of the life of verse than the bones of an animal reveal of its living grace'; 'the only recommendation for any mould is a perfect form produced from it'. On 30 Jan. Bayfield said: 'Prosody . . . provides the skeleton or framework on which verse is built'. The central issue, for the many letter writers was: how real or controlling is the metre?

his words. Thus Hardy read in Bridges: 'No line can be constructed without reference to its form: hence the same syllabic rhythms acquire different values according to their place in the line. The indefinable delicacy of this power over the hidden possibilities of speech is what most invites and rewards the artist in his technique.' ('Humdrum', 52.)

Though the Victorians did not answer some of the key questions about English metre, we should not underestimate the breakthrough in common sense about the subject which they achieved. At some point in reading the verse line, the mind perceives a principle of spacing—that within a numerical range of syllables there tends to occur a certain number of stresses. This perception can occur early in the reading of the poem, or late, particularly if the number of syllables and stresses vary a great deal. If this number varies beyond a certain point, the mind may never perceive the spacing. Once the mind perceives the principle of spacing, it can apply the principle to very diverse numbers of syllables and stresses. It can also perceive what configuration of syllables and stresses strains at the spacing, and what configuration violates the spacing.[38]

Hardy said (see below, Chapter 2) that he kept 'quantities of notes on rhythm and metre' which included 'verse skeletons . . . designated by the marks for long and short syllables, accentuations, etc.'. He seems here to associate metre with the designations (breves and longas) used for long and short syllables and traditionally adapted to English verse. The 'rhythm' then would seem to come from the interplay of the 'metre' and the speech rhythm, whose influence is denoted by 'accentuations'. Hardy's remark assumes that the metrical pattern and its variations can be pictured by dividing the line into 'feet'. There was much discussion on how suitable was the

[38] Patmore believed that rhyme helped establish the spacing within which the stipulated accents arranged themselves, and so made sense of the age-old issue of rhyme's relation to metre (*Essay*, 39–42). But see Omond's criticism, *English Metrists*, 175. Occasionally, Victorian writers will try to get at the notion of metrical awkwardness. For example, T. B. Rudmose-Brown writes: 'The line *is* in conformity with the time-scheme, unless its periodicity is obscured by important speech-accents in the interior of periods suggesting the *thinking* of the ictus at inappropriate points' (Omond and Rudmose-Brown, 'Is Verse a Trammel?', 549).

application of names of classical feet to English verse. Patmore
wrote:

The word *foot*, however, may be usefully retained in the criticism of
modern verse, inasmuch as it indicates a reality, though not exactly
that which is indicated by it with regard to classical metre. The true
meaning of the word for us is to be obtained from attending to its
employment by . . . musical writers, who speak of iambic, trochaic,
and dactylic *rhythms*. Thus, a strain in 'common time' beginning with
the unaccented note, is called iambic; a strain in 'triple time'
beginning with two unaccented notes, anapaestic, and so forth.
(*Essay*, 20–1.)

De Selincourt again made a reasonable summary of the issue:
'Whatever a foot may be, the time seems to have come when
prosodists should agree to accept the mere term as innocuous
and established by usage. The foot would then be to poetry
what the bar is to music—the rhythmical unit of verse.' (74)
The danger of the foot, an entity much maligned in modern
metrical criticism, is partly that it encourages the confusion of
classical quantities and English stresses, partly that it suggests
an unrealistic source of rhythm. It suggests that the rhythm is
built up out of separate segments like a chain built up out of
links. In fact, a foot is an abstract entity which results from the
metre's interaction with the given line. The metre stipulates
that the line fall into a set of spaces differentiated by a binary
pattern of accents and non-accents; and these spaces the
Victorians continued to call feet. The uniform spacing enables
us to picture differences in the rhythm of each line. This is what
Hardy is assuming when he says of his verse: 'He shaped his
poetry accordingly, introducing metrical pauses, and reversed
beats' (*Life*, 301).

Once acknowledgement of the difference between classical
quantities and English accentuation had been made, scansion
by feet, according to De Selincourt (93–4), was believed

required to state or exhibit two sets of facts; first, the rhythmical
abstract which a verse or stanza implies: secondly, the accentual or
syllabic liberties which appear in it. For the scansion of blank verse,
for instance, we must first lay down the scheme

$$\text{s \acute{s} | s \acute{s} | s \acute{s} | s \acute{s} | s \acute{s} |}$$

to indicate that we expect a line with five accents, dissyllabic feet, and a rising rhythm. Then, for the exhibition of the different formation of actual verses, we shall need three more symbols, ` for a displaced, x for a supernumerary, (´) for an implied accent; and we get results such as the following:

'Thùs at | their shád | y lódge | arríved, | bŏth stóod;

This statement from De Selincourt follows immediately after his statement about the mind hearing 'the implied accents in their places' as noted above. Hardy copied both statements into his notebook. De Selincourt's notion probably corresponds to Hardy's notion of a verse skeleton. If so, Hardy's verse skeleton reflected both the metrical norm and the different formations built thereon.

The Victorian question about the accuracy of foot scansion has often been raised since. As Wimsatt and Beardsley observe, this 'double character of scansion marks [to picture the influence both of the abstract pattern and objective linguistic features] has perhaps caused much of the difficulty in metrical theory' (99 n. 5). Alexander Ellis and Henry Sweet were famous for their complex discrimination of language features, a tradition of Victorian discernment emphasized by the linguists of the next century. Between picturing the simple metrical pattern, the complex language features, and their mutual influence, scansion achieves its rude compromise—which, if properly used, tradition sees as its virtue rather than its failure.

A given scansion is a rough sketch which may emphasize one of various aspects: the basic number of metrical spaces expected, the look of the stipulated feet, the rhythmic compromise, the linguistic material, or a given performance. These aspects are progressively more determined but less ruled: that is, we need the determination (some performance of the line) in order to find the rule; we need the rule in order to decide finally how to perform the line. Scansion intrudes and pictures some segment of the hermeneutical circle. The point of scansion is to help a reader hear the line's rhythm (at least as the critic hears it) and then note the basic pattern and its variation.

The tension in the 'double character' of scansion marks shows up in our uncertainty as to how to mark normally stressed syllables in unaccented positions ('| *both* stood |'), or normally unstressed syllables in accented positions ('Thùs *at* | their shá | dy lódge | . . .'), especially in the case of crux syllables, those normally unstressed syllables which must be accented if the metre is to be 'saved'.

Interestingly, we have examples of Hardy's own scansions. In his scansion of an hexameter line by Horace, he merely gives the spacing:

non ego | namque͡ para | bilem a | mo Vene | rem faci | lemque

In scanning Swinburne's 'sapphics', he uses longas and breves to suggest that the line is following a classical quantitative pattern:

Sāw thĕ whīte īmplācăblĕ Āphrŏdītĕ

(See below, Metrical Appendix.) Finally, in sorting out the rhythm of Shakespeare's rough iambic line beginning Act iv, scene iv, of *The Winter's Tale*, Hardy uses slash marks to indicate metrical accents:

These´ your unúsual weeds´ to each part´ of you´

In the minimal scansions done in this book, I have followed the traditional system, even though it is an anachronous combination of classical and accentual notations: breves for unaccented, slash marks for accented syllables, and reversed slash marks for what De Selincourt called 'supernumerary' accents. I have also used double slash marks for dipodic strong beats, and Hopkins's dot mark for a 'blank accent'. Such a dotted syllable, a sort of *stress potentia*, is the peg on which a new metrical norm can be constructed; such an accent may be actualized in order to maintain an intricate rhythm.

The Architectural Analogy

One of Hardy's most striking analogies for metrical form is that of Gothic architecture. In a major assessment of his metrical practice in the *Later Years*, he wrote:

In the reception of this and later volumes of Hardy's poems there was, he said, as regards form, the inevitable ascription to ignorance of what was really choice after full knowledge. That the author loved the art of concealing art was undiscerned. For instance, as to rhythm. Years earlier he had decided that too regular a beat was bad art. He had fortified himself in his opinion by thinking of the analogy of architecture, between which art and that of poetry he had discovered, to use his own words, that there existed a close and curious parallel, both arts, unlike some others, having to carry a rational content inside their artistic form. He knew that in architecture cunning irregularity is of enormous worth, and it is obvious that he carried on into his verse, perhaps in part unconsciously, the Gothic art-principle in which he had been trained—the principle of spontaneity, found in mouldings, tracery, and such like—resulting in the 'unforeseen' (as it has been called) character of his metres and stanzas, that of stress rather than of syllable, poetic texture rather than poetic veneer; the latter kind of thing, under the name of 'constructed ornament', being what he, in common with every Gothic student, had been taught to avoid as the plague. He shaped his poetry accordingly, introducing metrical pauses, and reversed beats; and found for his trouble that some particular line of a poem exemplifying this principle was greeted with a would-be jocular remark that such a line 'did not make for immortality'. The same critic might have gone to one of our cathedrals (to follow up the analogy of architecture), and on discovering that the carved leafage of some capital or spandrel in the best period of Gothic art strayed freakishly out of its bounds over the moulding, where by rule it had no business to be, or that the enrichments of a string-course were not accurately spaced; or that there was a sudden blank in a wall where a window was to be expected from formal measurement, have declared with equally merry conviction, 'This does not make for immortality'. (300-1)

Hardy thus 'fortified himself in his opinion [of the nature of metre] by thinking of the analogy of architecture', where 'cunning irregularity is of enormous worth'; 'he carried on into his verse . . . the Gothic art-principle . . . the principle of spontaneity, found in mouldings, tracery, and such like'. He also compared the way 'the carved leafage of some capital or spandrel in the best period of Gothic art strayed freakishly out of its bounds over the moulding'.

The analogy of Gothic architecture and metrical form plays

an interesting part in Victorian metrical theory. The analogy often lurks submerged in metrical discussion, for example the following by Robert Louis Stevenson. Stevenson has been discussing the counterpoint between groups of phrases and the metrical pattern:

The eccentric scansion of the groups is an adornment; but as soon as the original beat has been forgotten, they cease implicitly to be eccentric. Variety is what is sought; but if we destroy the original mould, one of the terms of this variety is lost, and we fall back on sameness. Thus, both as to the arithmetical measure of the verse, and the degree of regularity in scansion, we see the laws of prosody to have one common purpose: to keep alive the opposition of two schemes simultaneously followed; to keep them notably apart, though still coincident; and to balance them with such judicial nicety before the reader, that neither shall be unperceived and neither signally prevail. (555)

Issues of 'adornment' and 'the original mould' are issues Hardy would have discussed in the architect's office. The paradox of life and law applied to prosody found ready images in the tradition of Gothic architecture. (Again, Hegel may be the source of this analogy.)[39] Patmore's landmark article says:

The language should always seem to *feel*, though not to *suffer from* the bonds of verse. The very deformities produced, really or apparently, in the phraseology of a great poet, by the confinement of metre, are beautiful, exactly for the same reasons that in architecture justify the bossy Gothic foliage.

('The Gothic simile crushes me,' Ruskin wrote to Patmore in 1878.)[40]

[39] Though the connection between the subjects of metre and Gothic architecture is not developed by Hegel, the two subjects are discussed in similar ways by him. For Hegel, Gothic architecture is the third 'romantic' stage of architecture, following the symbolic (i.e. allegorical) and classical stages. Thus, for Hegel, Gothic architecture illustrates a dialectic of law and freedom, like that which the subject of metre illustrates.

[40] Patmore, *Essay*, 8 (Hardy copied a long section which comes 2 sentences after this statement by Patmore). Sister Mary Roth's footnote 8.24 on page 59 opened up the Gothic connection and led me to Ruskin's comments and Patmore's other essays. Ruskin's letter is in *Works*, xxxvii. 253. On the relation between Ruskin and Patmore's writings on Gothic architecture, see Michael Brooks, 'John Ruskin, Coventry Patmore, and the Nature of Gothic', *Victorian Periodicals Newsletter*, 12 (1979); 130–9.

Gothic architecture, as promoted by the Gothic Revival, was made by nineteenth-century critics a central example of organic or dialectic law. The eighteenth century by contrast invoked the analogy of classical architecture (Fussell, *Theory*, 47–9). The architectural symmetries of Gothic buildings, and the intricacy and 'spontaneity' of their details, seemed to illustrate dramatically the paradox of law and life. Gothic architecture was also used to illustrate the venerable Victorian principle of 'imperfect' art. Carlyle, Morris, Ruskin, Browning are a few names prominently associated with this principle. 'To banish imperfection,' said Ruskin in his essay on 'The Nature of Gothic', 'is to destroy expression, to check exertion, to paralyze vitality'.

One reason why Patmore's simile crushed Ruskin was that it was similar to one Ruskin had used earlier himself. The year before Patmore's article, Ruskin published the fourth volume of *Modern Painters* (1856). (Hardy listened to a colleague read extracts from *Modern Painters* in 1862—*Life*, 38.) In the fourth volume Ruskin wrote:

In vulgar ornamentation, entirely rigid laws of line are always observed; and the common Greek honeysuckle and other such formalisms are attractive to uneducated eyes, owing to their manifest compliance with the first conditions of unity and symmetry; being to really noble ornamentation what the sing-song of a bad reader of poetry, laying regular emphasis on every required syllable of every foot, is to the varied, irregular, unexpected, inimitable cadence of the voice of a person of sense and feeling reciting the same lines,—not incognizant of the rhythm, but delicately bending it to the expression of passion, and the natural sequence of thought. (*Works*, vi. 332–3.)

Ruskin's point is also like the one Wordsworth made in 1815 (cited above), but with the addition of a visual analogy, soon to be given an explicit Gothic context by Patmore. Ruskin does not invoke the Gothic analogy here and he probably did not

---

An image used by De Selincourt shows how foliage imagery and Gothic imagery co-operated: 'the art of the poet consists in introducing variation upon this basis of equality [the co-ordination of the syllables into feet and lines], in twining his works like vine-tendrils about the constant branches of the elm' (79). The bracketed phrase is by Hardy who copied the first half of this sentence (*Literary Notebooks*, ii. 209).

read Patmore's essay until it was published with *Amelia* in 1878, after which he acknowledged the 'crushing' power of the Gothic simile. In 1880, Ruskin would make the architectural ornament analogy central in his *Elements of English Prosody*. Here he discusses iambic metre and its 'perfect submission to dramatic accent'. In the dramatic pentameter, Ruskin wrote,

the divisions are purposefully inaccurate;—the accepted cadence of the metre being allowed only at intervals, and the prosody of every passionate line thrown into a disorder which is more lovely than any normal order, as the leaves of a living tree are more lovely than a formula honeysuckle ornament on a cornice;—the inner laws and native grace being all the more perfect in that they are less manifest. (372)

We can also trace the history of Patmore's use of the image. In 1849, he published an article reviewing Ruskin's *Seven Lamps of Architecture* and architectural works by Rickman, Durandus, and Brandon (a figure admired by Hardy). Patmore described how Gothic foliage imitates the way 'natural energy . . . develops and confines itself in regular forms'. Thus a spandrel from St. Alban's, 'a characteristic example of Perpendicular-Gothic leafage',

consists of a leaf or leaves and stalk, remotely resembling the vine, and adapting itself elegantly and exactly to the geometrical form of the spandrel, not, however, without apparently modifying that form, by prolonging the angle in the direction of the growth. Again, while the geometrical figure is modified by the energy of growth, the growth itself is modified and compressed by the geometrical outline; the leaf not only keeping that outline, but exhibiting that peculiar bulge or swelling, expression of compression from without, which approximates to deformity, but which is an inviolable characteristic of good Gothic leafage. ('Aesthetics', 65.)

In his 1850 review of *In Memoriam*, Patmore briefly noted: 'All beauty . . . from the "beauty of holiness" to the shape of the trefoil-clover leaf, is *life expressed in law*' (533). In his 1857 *Essay*, Patmore wrote: 'The co-ordination of life and law . . . determines the different degree and kinds of metre' (7).

In the same paragraph Patmore gives us his statement of the Gothic analogy which Ruskin found so crushing.

Hardy's first career as a creative artist was as a Gothic Revival architect (see Taylor, *Hardy's Poetry*, 48–59). The period in which he worked as an architect, 1856–72, represents the triumph both of the creative architect and of high Victorian Gothic. His *annus mirabilis* in architecture, when he won various prizes, was 1863. The imagery of Gothic patterns, with their '[p]etrified lacework—lightly lined / On ancient massiveness behind' ('The Abbey Mason', ii. 133), pervades Hardy's novels and poems in many subtle ways; among other things, it comes to symbolize patterns of human life, rigid and frail moulds into which lives and minds fall until they become a phantom of what they have been. The imagery, structure, atmosphere, and aesthetic issues of Gothic architecture nurtured Hardy's novels, poems, and finally his theory of metre and 'verse skeletons'.

When Hardy made the Gothic-metre analogy in the late 1920s, he was probably also influenced by an article he read and liked in 1919: Ramsay Traquair's 'Free Verse and the Parthenon' (*Canadian Bookman*, April 1919, pp. 22–6; *Collected Letters*, v. 305): 'Just as correct metre will not make a fine poem, so regular rhythm will not make a fine building'. On the other hand: 'Is not rhythm, regular rhythm, the very essence of poetry?'. The article explores the paradox of order and freedom, first in classical architecture ('no part can be taken away'), and then in Gothic architecture ('We may add a choir, aisles, chapels, cloisters, chantries, in what profusion we wish'). These two architectures are in turn related to strict and loose forms of rhythm in poetry.[41] 'The Parthenon has only one rhythmic form, and is a poem in a single metre, varied with the most exquisite skill without ever breaking the beat'; while 'our English cathedrals have many metres, each suited to its purpose'. In Lincoln Cathedral, for example, 'the complicated rhythm of clerestory window and vault' forms a complex counterpointing with the 'doubled beat of the triforium' below, and again with the 'steady slow beat of the nave arches' below

---

[41] Traquair is confusing 'free' verse with what Frost would later call 'loose iambics'.

that. 'As in simple music, the ornamentation, the rich melody is above, the rhythmic accompaniment is below.' Nevertheless, both architectures use 'old forms' within which variety is achieved. In this slight article, we can see how the tradition of Hegel mixing with Victorian metrical theory has become common currency—which Hardy draws upon as he thinks about the nature of his metrical rhythm.

One significance of the architectural analogy is that it is a visual analogy for an aural phenomenon. We have made much in this chapter of the 'abstract', i.e. unseen, nature of the metrical scheme; where it becomes visualized is in the printed lines on the page. The unheard melody is seen in the architecture of the verse stanza. To this theme, so fascinating to Hardy, we shall return in the last chapter.

2. *Hardy and Victorian Prosody*

'On Thursday, April 17, 1862', Hardy took the train to London (*Life*, 35). In so doing he brought his temperament and his Dorset experience into conjunction with a rich stream of poetic practice and theory. We have seen that the new era of Victorian metrical theory was inaugurated by Patmore's essay on 'English Metrical Critics' in 1857. In the 1860s followed a great revival and discussion of classical metres with their important relevance to English metrical issues. This decade also saw the beginning of landmark metrical discussions by Hopkins and his friends, and by scholars associated with the Philological Society. Out of this would eventually come the classics of the new prosody: books by Bridges, Saintsbury, Omond, and others.

The period of the 1860s also shows an extraordinary convergence of poetic and prosodic developments. The decade inaugurates the last great period in English of rhymed accentual-syllabic verse before the coming of a more eclectic modernism. As Hardy began writing, important developments were taking place in sonnets, ballads, hymns, quantitative classical imitations, French syllabic imitations, collections of verse, and theories of the relation of poetry and music.

The Poetic Tradition

In mainstream poetic practice, several important things were happening. Tennyson was at the height of his reputation. The Tennyson phenomenon (40,000 copies of *Enoch Arden* were sold in a few weeks in 1864, and 40,000 copies of the *Idylls* were pre-ordered in 1869) represents the climax of the English poet's relation to his public as prophet, teacher, and enter-

tainer. Never again would that degree of symbiosis be recovered. In this decade Browning finally recovered from his
reputation for *Sordello*, a work which seems to have influenced
Hardy now, and became a major figure in popular consciousness. His 1863 preface to *Sordello*—'my stress lay on the
incidents in the development of a soul', 'My own faults of
expression were many; but with care for a man or book such
would be surmounted'—always remained with Hardy, whose
prefaces to his own poems would later echo Browning's words.
The Pre-Raphaelites became widely influential now, especially
with the publication of Christina Rossetti's *Goblin Market* in
1862, Morris's *The Life and Death of Jason* in 1867, and *The
Earthly Paradise* in 1868–70.[1] But the great literary sensation,
one of the greatest in the history of English literature, was the
publication of Swinburne's *Poems and Ballads* in 1866, preceded
by *Atalanta in Calydon* in 1865. Swinburne's amazing new
rhythms complemented the new thoughts about prosody.
Hardy had been immersing himself in Tennyson and Browning, Shelley and Shakespeare, when *Poems and Ballads* burst on
the scene. The tercentenary of Shakespeare's birth in 1864 was
a kind of symbol of the prosperous literary experimentation of
the decade. If Hardy found himself in the midst of a revolution
of understanding in prosody, he also found himself in the midst
of a poetry fulfilling that prosody more elaborately than ever
before.

    Saintsbury's *History of English Prosody* is itself a good introduc-
tion to the setting of Hardy's poetry. Saintsbury traces the slow
growth of the art of accentual-syllabic rhythm out of the first
tentative steps in early English poets—'writing with two
entirely different systems in their ears' (i. 76). Saintsbury then
cites the first clear formulation of the rhythm in Chaucer, its
flourishing from Spenser through Shakespeare to Milton, its
retrenchment in the 'preceptist' metre of the Restoration and
eighteenth-century couplet, and its recovery of variety in the
romantic revolution. Saintsbury then celebrated Tennyson,

[1] On the 1860s as the great decade both of Pre-Raphaelite prestige and of Gothic
revival architecture, with their combined influence on Hardy, see Taylor, *Hardy's
Poetry*, 49.

Browning, and Swinburne as the poets who had climaxed the art of 'ordered liberty'. Saintsbury's accounts, necessarily oversimplified in his so-called 'Interchapters', may be a fair definition of the situation Hardy saw himself in when he embarked upon his 'new poetry'. (See below, Chapter 3.)

Saintsbury, also powerfully impressed by Swinburne in the 1860s, describes the poet as Hardy might have seen him. 'Every weapon', Saintsbury writes, 'and every slight of the English poet—equivalence and substitution, alternation and repetition, rhymes and rhymeless suspension of sound, volley and check of verse, stanza construction, line—and pause-moulding, foot-conjunction and contrast,—this poet knows and can use them all' (iii. 335). Citing Swinburne's knowledge of Elizabethan literature, classical literature, and French literature, Saintsbury adds: 'he possessed, as hardly anybody since Gray had possessed, the three arms—the horse, foot, and artillery— of classical, English, and foreign (not merely modern) verse and letters generally' (iii. 337). Saintsbury in particular singles out *Atalanta in Calydon* and *Poems and Ballads*.

This poetic prosperity of the 1860s reflects itself in other forms. We associate the sonnet primarily with the sixteenth century; but the Victorian age was the most prolific period for the writing of sonnet-sequences in the history of English literature. (The 1590s was the most prolific decade.) According to William Going, 234 sonnet sequences were published, accounting for over 4,000 sonnets. The most important example after Wordsworth's various series and after Elizabeth Barrett Browning's 'Sonnets from the Portuguese' (d. 1850) was George Meredith's *Modern Love* (d. 1862), followed by the first part of Rossetti's *The House of Life* in 1870. Robert Trewe, in Hardy's story 'An Imaginative Woman', seems to be modelled after Rossetti the sonneteer (see Rees). *Modern Love* was a collection of 16-line 'sonnets' (so called by Meredith and those who read him). It narrated the course of the death of a marriage from the husband's point of view. Shortly before his death in 1928, Hardy could still regret the hostile reviews which 'were strong enough to put a damper on the circulation of *Modern Love* till years after' (*Personal Writings*, 153). In the

sixties, Hardy's favourite poetic form was that of the sonnet.
Seventeen of the thirty-three poems dated before 1872 in the
Complete Poems are sonnets.[2] He would continue to write sonnets
until his last decade, for a total of thirty-eight sonnets. His
sequence, 'She to Him', of which we have four examples, is
'part of a much larger number which perished' (*Life*, 54).
These imitate Meredith's subject matter, in Hardy a tale of
unrequited love told from the woman's point of view. Interestingly Hopkins's career also began with sonnets in 1865 and his
sonnets were also about a spurned lover. The importance of
Swinburne and Meredith's early influence on Hardy can be
seen in the fact that an entire chapter of Hardy's autobiography will be entitled 'Deaths of Swinburne and Meredith,
1908–1909', all the other chapter titles being names of personal
events, places, or writings.

Equally important for Hardy was a major development at
this time in the history of hymns. *Hymns Ancient and Modern* had
been projected in 1858 and published in 1861. This hymnal
represented the climax of a long tradition which began with
Sternhold and Hopkins's 'Old Version' or sixteenth-century
version of the psalms (mostly in common metre), followed by
Tate and Brady's 'New Version' in 1696. The early use of these
simple forms and familiar psalms was reinforced by the practice
of 'lining out' or repeating lines sung first by the minister. Some
seventeenth-century hymns, most notably Bishop Ken's Morning and Evening Hymns were original compositions, and
composed for private devotions. In the eighteenth century
Isaac Watts, a Congregationalist, led the evangelical move
toward originally composed hymns that were not confined to
psalm translation and which were to be used in church. John
and Charles Wesley led a similar movement within the
established Church. As Watts said, the hymn '*should express the*

---

[2] On the sonnet, also see below, pp. 78–9 and Metrical Appendix. I have not
counted 'At a Lunar Eclipse' among the sonnets of the 1860s, where Purdy places it
(111). The source of Purdy's dating, in a 1918 Hardy letter to Gosse (*Collected Letters*, v.
246), is not conclusive; and other factors place the poem in the 1890s. I have discussed
this and other chronological issues in an unpublished essay, 'The Chronology of
Hardy's Poetry'.

thought and feeling of the singer, and not merely recall the circumstances or record the sentiment of David or Asaph or another'. Among many others, William Cowper and John Newton's influential *Olney Hymns* (d. 1782) emphasized personal introspection and set the standard for much of nineteenth-century evangelical hymnody. The spread of literacy in the nineteenth century enabled parishioners to read and sing a great variety of hymns. While hymns spread from chapel to church in the nineteenth century, nevertheless the High Church attitude toward hymn singing in church services was ambivalent and often hostile well into the nineteenth century. Thus John Keble's *The Christian Year* (1827) was intended not for church but for the 'soul at his bedside'. Mr Slope in *Barchester Towers* (d. 1857) considered hymns a sensuous Roman indulgence (chapter 6).[3]

Nevertheless, in the nineteenth century hymns were written by the tens of thousands and volumes published by the hundreds: 'They were sung almost everywhere, forming a truly popular culture of their own' (Tamke, 29). Gradually the simple traditional metres maintained up to about 1820 gave way to the more complex metrical forms of Keble and Newman. And hymns were no longer reserved for special occasions, holidays, and private devotions, but became a feature of High Church Sunday services. The generations of Hardys—including Hardy's grandfather, father, and uncle—who sang or played musical instruments in the choir at Stinsford church thrived on this development.

An important date is 1861 because it represents the definitive High Church acceptance of the practice of hymn singing in church. *Hymns Ancient and Modern* was not accepted everywhere, but it managed to lay down guidelines for the whole Church. The Hardy family had ceased all connection with the church choir by 1841 or 1842, but Hardy, who joined his father and uncle in playing the fiddle at parties and dances (*Life*, 32–3), was thoroughly immersed in hymnology. 'Though nominally

---

[3] On the history of hymnody, see Julian, Frost, and Tamke. The quotations from Watts and Keble are from Tamke, 20, 27.

unorthodox during the week Thomas was kept strictly at church on Sundays as usual, till he knew the Morning and Evening Services by heart including the rubrics, as well as large portions of the New Version of the Psalms.' 'We sang there in the good old High-And-Dry Church way—straight from the New Version' (*Life*, 18, 374). Hardy also loved to act out the church services at home (*Life*, 15). Though nostalgic always for the simple forms of Tate and Brady, whose hymns were often appended to church prayerbooks, Hardy was strongly influenced by the variety of forms and subjects in the new *Hymns Ancient and Modern*. In his own copy of the later 1889 edition of *Hymns Ancient and Modern*, I have noted markings by Hardy against 221 of the hymns. He also marked many of the hymns in his copy of Keble's *The Christian Year*, which he owned in an 1860 edition; he also owned an 1868 edition of Newman's *Verses on Various Occasions*. The Appendix at the end of this book makes clear the extent of the influences of these sources on Hardy.

In 1895 Hardy listed three hymns which 'have always been familiar and favorite hymns of mine as poetry: 1. Tate and Brady's "Thou turnest man, O Lord, to dust". Ps. xc. *vv.* 3, 4, 5, 6. (Tate and Brady.) 2. "Awake, my soul, and with the sun" (Morning Hymn, Ken.) 3. "Lead, kindly light" (Newman)' (*Life*, 275). Hardy's poem, 'Barthélémon at Vauxhall' (ii. 331) will celebrate Barthélémon's musical arrangement of Ken's hymn. These three hymns reflect the movement of hymnody from psalm to original composition to the complex personal hymnody of the mid-nineteenth century. One next step would be a 'hymn' like Hardy's 'The Impercipient' (i. 87) subtitled '(At a Cathedral Service)'. Hardy was more influenced by hymn (and ballad) forms than by any other metrical form. About 160 of Hardy's poems follow 'short', 'long', or 'common' measure, or one of the forms in *Hymns Ancient and Modern*. He continued to haunt church services throughout his life, and his works are filled with references to hymns and church choirs.

Complementary to the current developments in hymnology were important developments in the collection of ballads. The years 1857–9 saw the publication of the first version of Francis

Child's *English and Scottish Ballads*, organized according to subjects, and reissued in 1860, 1864, and 1866. This was the forerunner of the much more extensive version, *The English and Scottish Popular Ballads*, better organized according to metrical forms and published in ten parts between 1882 and 1898. Child did for ballads what the *Oxford English Dictionary* (which was inaugurated in 1860) did for words—recovered their original versions and traced their variants over the years (in so far as they could be dated).[4] In 1867–8, the *Bishop Percy Folio Manuscript*, edited by Hales and Furnivall, was published for the first time, giving Child more authentic texts for his 1882–98 work. The original Percy *Reliques* of 1765 was only a small and much corrected portion of this manuscript. The publication of the complete manuscript represented the most important publication in balladry since Scott's *Minstrelsy of the Scottish Border* in 1802–3. In 1868 the Ballad Society was founded. In 1869 the Roxburghe collection of broadsheet ballads was published. Hardy may have known William Allingham's popular selection of ballads, *The Ballad Book*, published in 1865. Hardy owned an 1857 edition of Percy's *Reliques* and an 187– edition of the *Ballad Minstrelsy of Scotland*.

Thus Hardy's interest in ballads was not that of a 'naif and rude bard who sings only because he must' (*Personal Writings*, 80). His balladry was consistent with his interest in other metrical forms which were undergoing interesting historical development or exploration at this time. The earlier ballad revival of the late eighteenth century had been of particular metrical interest because it had helped spread the use of 'real' trisyllabic feet into other genres of verse. The ballad had thus contributed to the development of the new prosody (Stewart, *Modern Metrical Techniques*, 47) whose principles were being formulated as the earlier ballads were finding their definitive scholarly forms.

At the end of the 1860s, another source of poetic forms began to come into vogue. In 1866 Swinburne published two

[4] On Hardy's relation to the *OED*, see Taylor, 'Victorian Philology'; on Hardy and ballads, see Taylor, *Hardy's Poetry*, 93–7. Hardy referred to Child in an 1889 letter (*Collected Letters*, i. 199).

rondeaux which he called rondels; and this marks a hint of the coming revival of the French lyric forms in the 1870s and following decades. The reappearance of these romance forms in English literature 'after the lapse of four centuries was due to several causes. . . . The forms found a much more general recognition at their second coming to England than they had in the age of Chaucer' (Cohen, 78). Their reappearance reflects a growing Victorian interest in complex stanza forms and intricate rhyme schemes. For a fifteen-year period beginning about 1868, rondeaux, villanelles, triolets, ballades, and other forms, were written and discovered. At the beginning of the 1890s, Florence Henniker gave Hardy a copy of Gleeson White's *Ballades and Rondeaus* (1887) (*Collected Letters*, ii. 24), which is a summary climax of this tradition: 'this collection is the first of its sort' (lv). Hardy would imitate several of its forms—the rondeau, triolet, Pantoum, villanelle—in various poems he wrote as he returned to a full-time career as a poet. White's introduction also reintroduced Hardy to the issue of the early romance syllabic influence on English verse.

The 1860s also saw the most influential of all Victorian anthologies of English poetry—Francis Palgrave's *The Golden Treasury* (1861). It represents a high Victorian assessment of the lyric accentual-syllabic tradition. Hardy owned and marked up a first edition given him by Horace Moule. It became his primary poetic manual. His only ambition, he said at the end of his life, had been to write a poem good enough to include in the *Golden Treasury* (*Life*, 444). Eventually Hardy's *Selected Poems* would be published as part of the Golden Treasury series. Palgrave's great anthology was a compendium of the major lyric forms in English, many of which Hardy would imitate in his own poems. Its early influence on Hardy can be seen in the use Hardy made of it in his first published novel, *Desperate Remedies* (1871). The allusions there to poems by Collins, Crashaw, Carew, Keats, Milton, Shelley, Wordsworth, and Spenser, are all allusions to selections in Palgrave.

Finally, the 1860s witnessed the greatest revival of imitation of classical metres since the late sixteenth century: 'criticism of prosody was then mainly concerned with these' (Omond,

English Metrists, 176). Arnold's *On Translating Homer* in 1860–1, Tennyson's and Swinburne's experiments, are perhaps the most prominent of a number of discussions and imitations, many of them concerned with hexameters. In fact, the swelling tide had begun in the 1840s with Longfellow, Clough, and much hexameter imitation and discussion. Omond's appendices show 18 'classical items' for the 1840s, then a lapse of 9 items for the 1850s, then a flood of 27 for the 1860s. Renewed discussion and experiments in classical verse made poets and theorists reconsider the puzzle of classical rhythm and its relation to the puzzle of English rhythm. Arnold argued for a compromise English imitation of Greek hexameters. He recommended that the translator make English accents correspond with Greek longs and avoid putting long English syllables in Greek short positions (194). He also recommended the translator avoid Longfellow's accentual dactylic monotony by following Clough's use of frequent spondaic substitutions (151)—a piece of advice that probably influenced Hopkins. Oh, to have been a fly on the wall during the conversation William Allingham recorded between Tennyson, Palgrave, and others in 1863: 'We talk of "Christabel". Race down, I get first to the stile. After dinner more talk of "Classic Metres".'[5]

The Metrical Appendix at the end of this book provides more detail on Hardy's relation to the hymn, ballad, sonnet, romance form, and classical traditions, as well as to the mainstream poetic tradition.

Hardy and the Prosodists

The 1860s was Hardy's most artistically formative decade. It is also the least well documented decade of Hardy's adult life. We know that he immersed himself in reading the mainstream

---

[5] Allingham, *Diary* (Carbondale, Ill., 1967), 94. Omond's appendices show: 1 classical item for the 1800s (but also include Coleridge's 'Metrical Experiments'), 2 for the 1810s, 3 for the 1820s including Southey's hexameter *Vision of Judgement*, 2 for the 1830s, 18 for the 1840s with Longfellow (of whom Hardy owned an 1860s edition), Clough, and much discussion of English hexameters, 9 for the 1850s, 27 for the 1860s, including again an enormous number of 'hexameter' experiments, 6 for the 1870s, 10 for the 1880s, 20 for the 1890s (including Hardy's 'The Temporary the All'), 20 for the 1900s, including many experiments by Bridges and reviews of these experiments.

poets of English literature. We shall see that he participated in the great classical experiments of the 1860s. His poems from this decade show the influence of contemporary sonnet, hymn, and ballad revivals. But evidence of his reading in the new prosodists only comes later. We have seen how Hardy took extensive notes about 1907 from an 1878 or later edition of Patmore's *Essay* and from De Selincourt's article of 1911. He owned and used *The Rhymer's Lexicon*, a 1905 edition with an introduction by Saintsbury summing up and previewing his forthcoming three-volume history. 'Any statement here made', Saintsbury noted, 'is ready to be justified in the proper place— an elaborate *History of English Prosody*' (v). This comment may then have led Hardy to the *History*: he was reputed to have read it in the British Museum when it was first published (Blunden, 119). Hardy also read and put in his *Literary Notebooks* (ii. 306) a cutting from a 29 June 1906 *TLS* review of Volume 1 of Saintsbury's *History of English Prosody*; the review concluded: 'To these volumes everyone properly interested in prosody will look forward with eagerness' (230). Hardy also includes in his *Literary Notebooks* (ii. 419–20) an extensive cutting from Basil De Selincourt's 1910 review of Saintsbury's *History* (*Outlook*, 13 August, 220). Hardy owned and marked various books which contained introductions by Saintsbury: *French Lyrics* (1883), Herrick's *Poetical Works* (1893), Donne's *Poems* (1896). In the latter he marked Saintsbury's comparison of Donne's prosody to that of the human satirists whose 'licenses' were deliberate variations from a 'correct' norm (p. xxi). On a late but undated list of books he intends to read, Hardy cites Omond's *English Metrists* (*Personal Notebooks*, 99). He also took notes from Bridges's 1922 essay on free verse (*Personal Notebooks*, 62–3). He owned a copy of Skeat's 1882 edition of Guest's *History of English Rhythms*. In the back of this volume, he would insert several clippings from the *TLS* discussions by D. S. MacColl and others in 1918–19. He owned a copy of Gummere's later work, *The Beginnings of Poetry* (1901). In 1919, the Canadian poet Alfred Gordon published an article entitled 'What is Poetry?—A Synthesis of Modern Criticism'. The article included a discussion of Saintsbury's and Mayor's notions of equivalence and substitution. Hardy read the article and wrote to Gordon. Gordon in reply sent Hardy another article called 'The New Prosody: A Rejoinder', published in 1920. Florence Hardy wrote to Gordon that Hardy planned to read this 'article on the New Prosody'. The phrase 'new prosody' here, however, refers to the strict musical scansionists of the school of Lanier; Gordon argued against them and in favour of Saintsbury.[6]

Hardy's first poems, then, were written in the aftermath of Patmore's article; his return to poetry in the 1890s occurred as Bridges, Saintsbury, and Omond were writing and revising their major works. One significance of Hardy as a poet is that he is perhaps the first major poet in the tradition of Sidney to know a theory of accentual-syllabic metre adequate to the richness of the verse.[7] The more Hardy realized the implications of these theories, the more his poetic practice and confidence improved. His earliest poems (those written before 1890) are in mainly conservative forms: sonnets (17 poems), quatrains (20 poems), and longer forms (21 poems), using either one or two different line lengths; also one example of blank verse, one attempt at Spenserian stanzas (no longer extant), two *terze rime*, and one traditional triplet.[8] It is after he

[6] For Gordon's letter 9 June 1920 ('you were so kind as to recognize my article, "What is Poetry?"') and Florence's reply, see Dorset County Museum letters, Folder *G General*. Gordon also sent Hardy his poems: see Maggs and Wreden sales catalogues. Hardy had a low view of Saintsbury as a general literary critic—prompted perhaps by Saintsbury's damaging review of *Tess*: see *Collected Letters*, i. 253, 255; Millgate, *Biography*, 321; D. F. Barber, ed., *Concerning Thomas Hardy* (London, 1968), 102.

[7] Of course the great English irony is that the Victorian formulation of the principles of accentual-syllabic verse lagged several hundred years behind the poets' accomplishments in the form. John Thompson has traced how the mature sense of the interplay of metrical form and language crystallized with Sidney. Hardy did not need the theory in order to practice the form. But knowing the theory helped confirm his intuitions about the form. In Lewes's *The Story of Goethe's Life*, Hardy read of Goethe: 'from Voss he tried to master the principles of Metre with the zeal of a philologist. There is something very piquant in the idea of the greatest poet of his nation, the most musical master of verse in all its possible forms, trying to acquire a theoretic knowledge of that which on instinct he did to perfection' (350).

[8] Three poems from this period use long stanzas with 3-line lengths. Of the 46 poems datable with reasonable probability to the 1860s, 36 use conventional or borrowed forms; only 10 forms are original to Hardy. 'At a Lunar Eclipse' is not counted in the sonnet statistics: see above, n. 2.

returns to poetry in the 1890s that he begins to reconsider the
new prosodists; and his verse forms become more complex and
interesting.

Hardy and Patmore

Toward the end of his life Hardy was ready to summarize for
himself his metrical principles. The summary occurs in the *Later
Years*, chapter 25, which was probably written in the late 1920s
(Purdy, 266). The chapter quotes several reflections Hardy
made in 1899 in response to some of the reviews of his first
volume of poems, *Wessex Poems*. The last reflection he quotes
from 1899 is: 'Poetry is emotion put into measure. The emotion
must come by nature, but the measure can be acquired by art'
(*Life*, 300). At this point in the manuscript Hardy intrudes his
major assessment, much of which we have quoted above
(pp. 42–3). We have seen how it echoes the metrical theorists'
use of the Gothic architecture analogy; it also contains several
other echoes of the metrical discussions we have been following.

Hardy's defence of his rough-hewn but cunning metres
sounds somewhat similar to the original conclusion of
Patmore's *Essay*:

Most readers of poetry, and we fear we must add, modern writers
upon it, know nothing, and feel nothing, of the laws of metre as they
have been practiced by all great poets. 'Smoothness' is regarded as
the highest praise of versification, whereas it is about the lowest and
most easily attainable of all its qualities. . . . [To] speak of 'smoothness' as anything more than the negative, merely mechanical and
meanest merit of verse, is to indicate a great insensibility to the nature
of music in language. Such insensibility is, however, the almost
inevitable result upon most minds of the unleisurely habits of reading
into which we moderns are falling. We have not time to feel with a
good poet thoroughly enough to catch his music, and the consequence
is, that good poets have lately been writing down to our incapacity.

There is much to suggest that Hardy may have known of
Patmore's theory of prosody long before he took notes in the
1900s. Hardy owned *The Angel in the House* in a two-volume
1860–3 edition given him by Patmore in 1875. That year,

Patmore wrote to Hardy to express regret that the beauty and power of *A Pair of Blue Eyes* 'should not have assured themselves the immortality which would have been impressed upon them by the form of verse' (*Life*, 104–5). Hardy was 'much struck by this opinion from Patmore' and partly as a result began a careful study of prose style. At this time he made several notations on style, none of them of great profundity. But these 1875 notes were perhaps influenced by Patmore, and we can see their development into the *Later Years* assessment. The notes read:

'Am more and more confirmed in an idea I have long held, as a matter of common sense, long before I thought of any old aphorism bearing on the subject: 'Ars est celare artem'. The whole secret of a living style and the difference between it and a dead style, lies in not having too much style—being, in fact, a little careless, or rather seeming to be, here and there. It brings wonderful life into the writing:

> 'A sweet disorder in the dress . . .
> A careless shoe-string, in whose tie
> I see a wild civility,
> Do more bewitch me than when art
> Is too precise in every part.

'Otherwise your style is like worn half-pence—all the fresh images rounded off by rubbing, and no crispness or movement at all.'

It is, of course, simply a carrying into prose the knowledge I have acquired in poetry—that inexact rhymes and rhythms now and then are far more pleasing than correct ones. (*Life*, 105.)

When Hardy says in his *Later Years* assessment, 'Years earlier he had decided that too regular a beat was bad art', he seems to be referring to this 1875 passage. In 1886 Hardy corresponded with Patmore concerning Patmore's article on Barnes: 'Your criticism . . . was most instructive' (*Collected Letters*, i. 157). In 1890 Hardy read Patmore's *Principle in Art* (which includes discussion of pathos in some Hardy novels) and wrote to the author: 'I have read some of the Essays: and am now reading two or three every evening immediately before bedtime: to my profit I hope' (*Collected Letters*, i. 208). In his *Literary Notebooks* (ii. 103), around 1900, he quoted some of Patmore's opinions.

In 1899, then, Hardy returned to these themes as we have seen. Interestingly, in this same year, he corresponded with Mrs Coventry Patmore who told him of Coventry's admiration for the 'consummate art and pathos' of *A Pair of Blue Eyes* (*Life*, 302). At the beginning of 1900 Hardy was thinking about Wordsworth's argument in the appendix to *Lyrical Ballads* and noted its 'confusion of thought' (*Life*, 306). To clarify the issue for himself, he drew a diagram illustrating the overlapping relation between 'works of *passion and sentiment*' and 'works of *fancy* (or *imagination*)': in the former 'the language of verse is the language of prose'; in the latter ' "poetic diction" (of the real kind) is proper, and even necessary'. Hardy may have been influenced by a somewhat parallel discussion in Patmore's *Essay*. At a later time he would copy from Patmore: 'Wordsworth's erroneous critical views of the necess[ity] of approx[imating] the lang[uage] of poetry, as much as possible, to that of prose, especially by the avoidance of gramm[atical] inversions, arose from his having overlooked the necessity of manifesting, as well as moving in, the bonds of verse.' (*Literary Notebooks*, ii. 190; Patmore, *Essay*, 9.) Hardy will include in his *Literary Notebooks* (ii. 409) a sentence from a 15 March 1905 *TLS* review of Gosse's biography of Patmore. The review discusses the 'simple regular metre . . . symbolic of its theme' in 'The Angel in the House', and contrasts it with the 'most irregular' metre of the later 'The Unknown Eros'. The review also notes how Patmore's 'originality gained power from obedience to law', echoing language in Patmore's great essay. In 1913 Hardy will refer on two occasions to Patmore's admiration of *A Pair of Blue Eyes* (*Collected Letters*, iv. 288, 291). Again in 1918 Hardy will refer to Patmore in an essay on Barnes where Hardy remembers Patmore's high opinion of Barnes's 'poetry of profound art' (*Personal Writings*, 84). Finally, in his *Personal Notebooks* (47), Hardy will quote from a 1921 *TLS* article on Patmore's literary criticism.

When, therefore, about 1907 Hardy took extensive notes from Patmore's 1906 edition of the *Essay*, he may have been rereading an essay he had considered long before. We know other examples of Hardy later taking notes from essays which he had read years before.[9] By the time Hardy took notes from Patmore, his renewed poetic career was well under way and he was ready to make a decisive turning toward more complex and inventive metrical forms, as we shall see. He was ready then to define more explicitly for himself the principles of the new prosody.

The *Later Years* assessment is sometimes assumed to show that Hardy thinks of his poetry as being a sort of 'accentualist' verse measured by number of accents alone. Hardy cites the ' "unforeseen" (as it has been called) character of his metres and stanzas, that of stress rather than of syllable, poetic texture rather than poetic veneer'. But close analysis of the passage, plus consideration of the sources from which Hardy draws, clearly places it with Victorian new prosody and the accentual-syllabic tradition. Interestingly the *Later Years* assessment begins in a manner similar to the way De Selincourt's article ends: 'For the rider of a fiery horse soon learns that he must govern its paces if he is not to be governed by them. "Ars est celare artem." Accent, at least, serves the artist best when it is concealed most artfully' (96). So the *Later Years* assessment says: 'That the author loved the art of concealing art was undiscerned'. We have seen Hardy quote the Latin formula in 1875 and he would allude to it again in the *Life* (384). In 1919, he expressed some advice for a young poet: 'Be also very careful about the mechanical part of your verse—rhythms, rhymes, &c. They [critics] do not know that dissonances, and other irregularities can be produced advisedly, as art, and worked as to give more charm than strict conformities, to the mind and ear of those trained and steeped in poetry . . . *Ars est celare artem*' (*Personal Notebooks*, 272). This classical formula was associated in Victorian times with the emerging notion of counterpoint. Hunt, writing in 1844 on Coleridge's *Christabel*, uses a related dictum: 'The principle of Variety in Uniformity is here worked out in a style "beyond the reach of art"' (58). The Hunt dictum is appropriately Romantic, Longinian. Hardy and De

---

[9] On Hardy's later notes from Arnold, see *Literary Notebooks*, i. 344–5, 363–4, 393; on Hardy's later notes from Wordsworth's preface, see Purdy, 113, Wright, 18, *Life*, 306.

Selincourt's dictum is more consistent with new prosody principles whereby the 'predetermined score' is 'concealed' by the overlay of seemingly spontaneous speech rhythms. This concealment is what Hardy calls 'cunning irregularity'.

The ' "unforeseen" (as it has been called) character of his metres and stanzas' is therefore consistent with the idea of a verse skeleton in which they are anchored. Thus Hardy 'shaped his poetry accordingly, introducing metrical pauses, and reversed beats'—a comment which only makes sense in terms of a metrical norm. The new prosody confirmed Hardy in his view that it was possible to reject 'constructed ornament' like eighteenth-century metrical conservatism and achieve an art of concealing art. Hardy's other analogies also derive from the new prosody. 'The enrichments of a string-course . . . not accurately spaced' acknowledges the musical analogy, so important for Omond and others, without applying the precision of musical bars. 'A sudden blank in a wall' is felt only because there 'a window was to be *expected* from formal measurement'. We have seen that expectation was a key concept of the new prosody. When Hardy uses the term 'poetic texture', we may remember that Stevenson had used this term to denote the speech rhythms as they are counterpointed with the metrical scheme. Hardy's distinction between 'rhythm and metre' in the last paragraph of the assessment seems to confirm his awareness of these issues.[10]

In this remarkable assessment, Hardy goes on to say:

If any proof were wanted that Hardy was not at this time and later the apprentice at verse that he was supposed to be, it could be found

---

[10] Hardy's description of William Barnes's metres is consistent with Victorian notions of metrical expectation: 'Barnes . . . really belonged to the literary school of such poets as Tennyson, Gray, and Collins, rather than to that of the old unpremeditating singers in dialect. Primarily spontaneous, he was academic close after; and we find him warbling his native wood-notes with a watchful eye on the predetermined score, a far remove from the popular impression of him as the naif and rude bard who sings only because he must, and who submits the uncouth lines of his page to us without knowing how they come there.' (*Personal Writings*, 79–80.) Hardy goes on to praise Barnes for his 'careful finish . . . verbal dexterities' and 'searchings for the most cunning syllables'. By 'a felicitous instinct he does at times break into sudden irregularities in the midst of his subtle rhythms and measures' (*Personal Writings*, 80–1).

in an examination of his studies over many years. Among his papers were quantities of notes on rhythm and metre: with outlines and experiments in innumerable original measures, some of which he adopted from time to time. These verse skeletons were mostly blank, and only designated by the usual marks for long and short syllables, accentuations, etc., but they were occasionally made up of 'nonsense verses'—such as, he said, were written when he was a boy by students of Latin prosody with the aid of a 'Gradus'. (*Life*, 300–2.)

In using the term 'verse skeleton', Hardy is echoing its deliberate use by Bridges and Saintsbury, who in turn may be echoing Thelwall's interesting remark about the 'skeleton rhythmus' cited above—though Thelwall did not invest the term with the aura of abstraction which it later took on. In *The Philology of the English Tongue* (1887 edition), John Earle noted: 'Metre is to rhythm what logic is to rhetoric; what the bone frame of an animal is to its living form and movements' (631). In his *Practical Hints on the Quantitative Pronunciation of Latin* (1874), Ellis begins his analysis of feet with the statement: 'Hence we reduce a verse to a mere skeleton of sound, independent of sense and of rhetorical alterations of sound. These are the muscles and nerves to be laid on afterwards' (27). Bridges said that the 'accentual speech-rhythms which overlaid the metre' 'were like the flesh on a skeleton, and it was one advantage of this system that the skeleton gave free play to the flesh'. Saintsbury discusses the accentualist prosodists who count only stresses and says such a prosodist 'confines himself to the mere skeleton of the lines, and neglects their delicately formed and softly coloured flesh and members . . . some . . . seem to regard the stresses of a whole passage as supplying . . . a sufficient rhythmical skeleton the flesh of which—the unaccented or unstressed part—is allowed to huddle itself on and shuffle itself along as it pleases'. T. S. Eliot will adapt the image in a famous passage: 'the ghost of some simple metre should lurk behind the arras in even the "freest" verse'.[11]

---

[11] Bridges, *Milton's Prosody* (1921 edn.), 85; Saintsbury, *Historical Manual*, 10, 12; Eliot, 'Reflections on *Vers Libre*' (d. 1917, in *To Criticize the Critic* (New York, 1965), 187). Bridges: 'In both the quantitative and syllabic systems of verse there were strict

The fact that for *Paradise Lost* Bridges sees the 'skeleton' in the stipulated number of syllables of English verse (and in the quantities of classical Latin verse), while Saintsbury sees the 'skeleton' in the stipulated pattern of accents, is indicative of the difficulty of determining what remains constant (or skeletal) and what is varied. Hardy's use of the term 'verse skeleton' was perhaps influenced also by the skeletal outline of verse forms contained in the introduction to some graduses. (For examples, see Metrical Appendix, below.)

Hardy's reference to the gradus is to the *Gradus ad Parnassum*, a dictionary of synonyms, word usages, epithets, and most importantly lines of Latin verse with each of the syllables scanned. Many of the lines were typical examples from classical verse forms, like the dactylic hexameter. The schoolboy, told to write a Latin verse with the correct scansion, would use the gradus and select those words and lines with the needed sets of shorts and longs. If for most schoolboys the gradus promoted a 'brick-puzzle conception of the way to piece together Latin hexameters and pentameters' (Archer, 86), for a few school-boys, who were potentially serious poets, the exercises were valuable. Shelley, the greatest specific poetic influence on Hardy after Swinburne, and the master poet of English verse music, was an accomplished maker of Latin verses and gradus exercises in school. The biographies of Shelley, which Hardy owned, refer to the phenomenon (i.e. Jeaffreson, i. 48, 57). Also of interest is that some of Shelley's mature poetry, including a Hardy favourite (*Personal Writings*, 107) and Palgrave selection, 'O World!, O Life!, O Time!', was apparently first composed in the form of nonsense verses—a fact revealed in Buxton Forman's 1911 edition of Shelley's *Note Books* (i. 198–200).

Use of the gradus, instead of encouraging rigid schemes,

---

syllabic rules which gave the metre, while the accentual speech-rhythms . . . were like the flesh on the skeleton', etc. For Bridges, the 'skeleton' was lost as later writers imitated only the accentual rhythm. In an early (1901) edition of *Milton's Prosody*, however, Bridges used 'skeleton scheme' in a way consistent with Saintsbury. Bridges cites in Shelley's 'Away, the Moor Was Dark' 'the various stress-rhythms with which he was, so to speak, counterpointing the original measure' (104) and notes 'the skeleton scheme of four double stresses and the break in the middle'. Hardy composed his passage on 'verse skeletons', we remember, probably some time after 1925 (Purdy, 266).

would encourage a sense of this mystery of interplay—for here the puzzlement of the average schoolboy is consistent with the insight of a Shelley. No Englishman knew very confidently what those classical quantities sounded like. To make a proper verse he had to follow blindly such rules as 'syllables are long if they consist of long vowels, or vowels before double consonants, with certain exceptions and additions'. Ellis in 1874 described the state of Latin verse reading for the student at Eton not long after Shelley had left: 'if in his verses he followed the laws of his "gradus" which were at utter variance with the custom of his speech, he was . . . held to have made no "false quantities". That he did not pronounce a single vowel correctly by intention, that he did not understand the nature of long and short vowels or syllables, or the rhythm they made in verse (except as by "gradus" aforesaid), that he had no conception of what the nature of Latin accent was, and that Latin as he uttered it (not as he saw it) was pure *vōx et praeterēa nihil*, sound without any sense at all to a Roman's ears—of this he had no conception whatever.'[12]

Some students in English classes today cannot understand English accent, but they might be able to write an iambic pentameter by merely following the dictionary marks for stressed syllables. Such was the common experience of the nineteenth-century English schoolboy in regard to Latin verse. Even the Renaissance imitators in English of classical quantities seemed to make a distinction between the way the verses sounded and the way they scanned according to somewhat abstruse rules. If the rules were clear but the rhythm obscure for English readers of classical verse, the rhythm was clear but the rules obscure for their reading of their native verse. The whole subject of metre, for poets trained in this system, had an aura of intriguing mystery.

Hardy's remark about 'verse skeletons . . . designed by the usual marks for long and short syllables, accentuations, etc.' seems to refer to the Latin scansion marks, longas and breves,

---

[12] The Ellis quotation is from *Practical Hints* viii, partly quoted by Attridge, *Well-weighed Syllables*, 67; also see Hendrickson, McKerrow.

with 'accentuation' denoting the speech stresses. How this applies to the scansion of an English line, in Hardy's mind, is an interesting question. The adaptation of a Latin verse skeleton to an English verse skeleton returns us to an issue of great interest to the new prosodists and the classical experimenters of the 1860s: namely the relation between Latin and English rhythm.

Hardy's distinction between 'long and short syllables' and accentuation suggests that he may have been aware of the unsettled relation between quantities and accents in Latin verse. In *On Translating Homer* Arnold had complained: 'It must be remembered . . . that, if we disregard quantity too much in constructing English hexameters, we also disregard accent too much in reading Greek hexameters' (162 n. 1). In his *History of English Sounds* (1888), Henry Sweet would summarize: 'Even in Greek there can be no doubt that the natural quantities were often forced in metre; and in English the best poets are influenced by an unconscious respect for the natural quantities of the language' (103). There had been a general consensus— though disputed to the present day—that Latin verse had taken its metrical rules of quantity from the Greek, but had kept its native accents. The relation of the accents to the rule-governed quantities was not entirely known. In most cases, perhaps, the accents were left to take care of themselves while the quantities were strictly observed. It is possible, however, that some sort of conscious relation existed, either by which accents in certain positions deliberately coincided with long syllables (as seems to be the case in the last two feet of Virgil's hexameter) or deliberately conflicted (for example near the caesura in the hexameter). At any rate, the educated reader might note a kind of interplay between accents and quantities. De Selincourt argued that in Latin verse, 'it is from this clash of the arrhythmical [*sic*] against the rhythmical features of a language, of the elements which have rejected measurement against those which have submitted to it, that there arises the music of its verse' (80). The point was well known to classical scholars and perhaps to a few schoolboys like Hardy and Hopkins not overwhelmed by the technicalities of the gradus.

One interesting perception, which seems to have crystallized with the Victorian period, was expressed by Longmuir, as Hardy read (in the introduction to Walker's *Rhyming Dictionary*):

Those who have treated this subject have been led to point out the distinction between *quantity* and *accent*, affirming that the former regulated the versification of ancient, and the latter that of modern times. We humbly conceive that quantity and accent are found in the compositions of both periods; but that the former were more distinguished by the quantity of syllables, not neglecting the accent, and that the latter are more dependent on accent, not however, neglecting quantity. (xi–xii)

William Johnson Stone had noted this inverse parallel in his influential 1899 work *On the Use of Classical Metres in English* (6), reprinted with Bridges's *Milton's Prosody* in 1901. De Selincourt's summary seems generally correct and generally accepted in the Victorian period: 'the music of English, like that of classical, verse depends to a great extent on the relation of the accents to the quantities, only that that relationship is reversed' (95). In one of the *TLS* letters (23 January 1919) which Hardy put in the back of his copy of Guest, Adrian Collins explains this inverse relation of quantities and accents in Latin and English (p. 45); and on 6 February, D. S. MacColl replies: 'Mr. Collins' interesting letter brings good news of the spread of the new doctrines in the schools . . . I should be very curious to hear schoolboys reproduce the delicate balance of speech-accent and metrical ictus in Latin verse that recent theory propounds' (69).

Because of his background in Latin prosody, Hardy may have felt such interplay at work in much standard English verse. Only here the interplay is one between the pattern of accents, which is metrically controlled, and the quantities, which play more freely. This kind of interplay constitutes an addition to that which we have already discussed, that is, where the interplay occurs within accentuation itself, between the expected and the actual configuration of accents and non-accents. While committed to this essential counterpoint (in

Stevenson's sense), Hardy also may have been interested in the effect of quantity on the rhythmic movement of the verse, much the way Hopkins was. We shall see that Hardy made extensive classifications of sound values and compiled notes on phonetics. But neither Hardy, nor Victorian theorists, were able to formulate clearly the relation of English quantities to the accentual metre. This has remained an unresolved problem.

Interestingly, the gradus exercises provided an analogy for exercises with rhyme and rhythm in English verse. John Longmuir made the analogy in his 1865 introduction to an important technical manual which Hardy used often in writing poetry: John Walker's *Rhyming Dictionary*, which Hardy owned and extensively annotated. In Longmuir's introduction, he read:

Whether the *Rhyming Dictionary* forms an indispensable part of the apparatus of the writing desk of the English poet, we have no means of ascertaining; but, if any good effect is produced by the composition of 'nonsense verses' in our Latin seminaries, we are persuaded that no small benefit would result from the practice of composing sense verses in our English schools; an ear for the pleasing rhythmus of the language would be cultivated, if not produced; the proper accentuation of the words would be fixed in the memory; the correct sound of the vowels in the rhymes would be learned; the discernment of nearly synonymous words would be forced on the attention, and the pupil's vocabulary would necessarily be greatly increased from the obligation that the measure of the verse would lay upon him to try several words before he could find one that would suit its place; not to speak of the due cultivation of the imagination, the elevation of the mind, and the refinement of the heart. (vii)

Hardy's English Stanza Forms

In the light of Victorian prosody, both theory and practice, Hardy's fascination with English accentual-syllabic metrical forms is easier to comprehend. His metrical forms constitute a fascinating treasury of technical structures. 'No English poet,' Auden said, 'not even Donne or Browning, employed so many and so complicated stanza forms. Anyone who imitates his style will learn at least one thing, how to make words fit into a complicated structure and also, if he is sensitive to such things, much about the influence of form upon content.' For some 1,093 poems, Hardy used well over 790 different metrical forms, that is, forms distinguishable by rhyme scheme and number of accents per line. (This does not include different stanza forms within the same poem.) By comparison, according to one estimate, Swinburne used 420, Tennyson 240, and Browning 200. As far as I can see, Hardy invented more verse forms than any other poet in the accentual-syllabic tradition: well over 620 verse forms in which almost 600 of his poems are written. At the same time, his poetry is full of scores of echoes of traditional forms or forms created by earlier poets. He imitated at least 170 such forms in which almost 500 poems are written, or partly written. A glossary of Hardy's metrical borrowings, which I have included as an appendix, serves as a comprehens- ive guide to the English metrical tradition.[13]

The *Life* records that on several occasions Hardy made an intensive study of lyric poets. For two years, 1866–7, he read no books of fiction or prose—believing 'that, as in verse was concentrated the essence of all imaginative and emotional literature, to read verse and nothing else was the shortest way to the fountain-head of such' (*Life*, 48). At the time of his Italian trip in 1887, a trip which signalled the beginning of his gradual 'return' to poetry, he again read widely in poetry (*Life*,

[13] Auden, 'Literary Transference'; Robert Fletcher, 'The Metrical Forms Used by Certain Victorian Poets', 87–91. Elizabeth Hickson, *The Versification of Thomas Hardy*, categorizes each of Hardy's poems (from the 1924 *Collected Poems* and the 1928 *Winter Words*) according to rhyme scheme, number of stresses per line, and rhythm (rising, falling, duple, triple, and combinations of these). She bases her nomenclature on T. S. Omond's *Study of Metre*. Various of Hickson's readings of individual lines are incorrect (as might be expected in a work of such incessant detail), but her basic outline and general work is very valuable, indeed indispensable to my own research: see her 'Appendix. Table VI. Stanza-Patterns of Thomas Hardy's Lyrical and Narrative Poems', 92–119. In the figures I have given, I have not considered differences in rhythm (i.e. falling versus rising), or different stanza forms within the same poem. Hardy's estimated 799 metrical forms include 742 in the *Collected Poetical Works*, 41 in the *Dynasts*, 8 in the *Queen of Cornwall*, and 8 in the uncollected poems. Hardy's poems using imitated forms (not including his imitations of earlier forms he invented) include approximately 452 poems from the *Collected Poetical Works*, and 46 poems from the *Dynasts*, *Queen of Cornwall*, and uncollected poems. His other 595 poems are in forms he invented. 'Imitated forms' here means forms listed in the Metrical Appendix.

203). He continually reread his poets and knew them so well that in later editions and collections he corrected misedited lines. Asked in 1910 about modern poetry, he said that 'he saw most of what appeared in the literary reviews and magazines and, if he liked what he read, would procure the collected work of a poet' (Hedgcock, 291). Also, he noted: 'I have been lead to read poets by seeing specimens in these anthologies' (*Collected Letters*, iv. 256); the anthologies which most influenced him are described in the Metrical Appendix. When he read other poets, he was extremely interested in the metrical form they used and often paid them the compliment of borrowing it. These compliments were paid both to the famous and the obscure. As a part of the advice he proposed giving a young poet in 1919 (see above), he said: 'The only practical advice I can give, and I give that with great diffidence, is to begin with *imitative* poetry, adopting the manner and views of any recent poet' (*Personal Notebooks*, 272). In the case of old songs, he imitated their metrical forms as a way of preserving them because they were fast fading from popular consciousness, 'the orally transmitted ditties of centuries being slain at a stroke by the London comic songs' (*Life*, 20).

An early passage in the *Life* (16) shows that Hardy was keenly sensitive to resemblances in musical forms:

> When but little older he was puzzled by what seemed to him a resemblance between two marches of totally opposite sentiments— 'See the conquering hero comes' and 'The Dead March in *Saul*'. Some dozen years were to pass before he discovered that they were by the same composer.
>
> It may be added here that this sensitiveness to melody, though he was no skilled musician, remained with him through life.

In a manuscript version of the *Life* (*Life and Work*, 21), the last sentence is much longer: 'It may be added here that this sensitiveness to melody which, though he was no skilled musician, remained with him through life, was remarkable as being a characteristic of one whose critics in after years were never tired of repeating of his verses that they revealed an ear deaf to music.' The connection between the published and unpublished passages suggest that Hardy's sensitivity to musical resemblances has an important bearing on the metres of his poems, whose echoes the critics might not perceive.

For Hardy the specific difference between poetry and prose was metrical form. When he adapted a verse form he did so—as he said in the case of Latin hymns—that 'English prosody might be enriched' (*Life*, 306). In reading the many contemporary volumes of poetry that were sent to Hardy, one is increasingly impressed with the fact that traditional metrical form still was a vital 'heuristic' resource for Hardy and his contemporaries, both Victorian and Georgian. There was some fear, expressed by Bridges for example, that rhyming accentual-syllabic stanzas had exhausted their potential.[14] But the multiple ways in which metre could interact with meaning, and together interact with the tradition, was a central inspiration for Hardy and the poets he followed.

It is this dimension which Hardy seems to have in mind when he writes to Gosse in 1883: 'Lodge's poem to Rosaline was one of the first two or three which awakened in me a true, or mature, consciousness of what poetry consists in—after a Dark Age of five or six years which followed that vague sense, in childhood, of the charms of verse that most young people experience.' 'Rosaline' appears in Palgrave's *Golden Treasury*, Hardy's favourite anthology of the 1860s, and is a fairly complex ten-line stanza. About two years after writing to Gosse, Hardy thanked Ward and Downey for sending him *Songs from the Novelists* (*Collected Letters*, i. 135). The volume contains another poem by Lodge, 'Rosalynd's Madrigal', which seems to be the model for Hardy's poem, 'Weathers', a form in turn imitated by Anthony Hecht in 'Spring for Thomas Hardy'. Exploration leads to inspiration and imitation. A notebook by Hardy entitled 'Poetical Matter' contains the interesting note: '*Lyrical Meth* Find a situn from expce. Turn to Lycs for a form of

---

[14] For example, Bridges, 'Letter' (revised version), 71; Bridges, 'Wordsworth', 30–1. Donald Wesling sees Hardy as a victim of the exhaustion of traditional metrical forms in the Victorian period: see 'The Prosodies of Free Verse', 158. Elizabeth Schneider (*Dragon*, 44–8) also discusses the Victorian fear of metrical exhaustion.

expressn that has been used for a quite difft situn. Use it (same sitn from experience may be sung in sevl forms)'.[15]

Hardy may also have been influenced by a review he copied in his *Literary Notebooks* (ii. 239). The review, 'Mr. Swinburne as a Master of Metre', *Spectator*, 102 (17 April 1909), 605–6, was critical of Swinburne's habit of metrical borrowing, but fascinated by it as well. We see how 'Mr. Swinburne had taken up and developed a metre that had charmed him in the work of a poet whom he so greatly admired': i.e. Landor's 'If you no longer love me' behind Swinburne's 'Rococo'. (This parallel, like others the review cites, is much less exact than the kind listed below in the Metrical Appendix.) 'Like a scientific gardener, he took the flower, transplanted it, crossed it, and developed it.' Hardy copied the review's initial conclusion: 'his metrical effects were produced by a conscious, or even a mechanical, process. . . . That is why his verse . . . in the end is apt to weary'. But the review tends to belie this conclusion by the extent to which it acknowledges Swinburne's 'wonderful science': 'Though we have pointed out that Mr. Swinburne was to some extent a "nurseryman" in the world of poetry, it must not be supposed that we fail to realise the wonderful science of his achievement. . . . Metrical triumphs such as his . . . will survive as long as the English language.' Hardy will also become a 'scientific gardener'. The interplay between original and traditional forms is a constant factor in Hardy's poems. A pre-eminent example is, appropriately, Hardy's poetic tribute to Swinburne, 'A Singer Asleep'. The poem is a stanzaic *tour de force*, in which a series of eight original stanzas, all different, *abbaa*, *abbacc*, and so on, are made to issue into the last stanza, an *ottava rima*, *ababababcc* (with a trimeter last line, however); thus the poem seems to recapitulate the phylogeny of a traditional stanza out of the evolutionary fragments of original forms.

Indeed, for every stanza form Hardy borrowed, he invented almost four. In 1879 he quoted a sentence from Leslie Stephen:

[15] *Collected Letters*, i. 135; Hecht in *Poetry*, 82 (April 1953), 17; Millgate, citing the 'Poetical Matter' notebook in *Biography*, 89.

'The ultimate aim of the poet should be to touch our hearts by showing his own, and not to exhibit his learning, or his fine taste, or his skill in mimicking the notes of his predecessors.' The *Life* adds: 'That Hardy adhered pretty closely to this principle when he resumed the writing of poetry can hardly be denied' (128). In 1907, Hardy told an aspiring poet that he particularly enjoyed looking at her poems because 'their form seems to insist upon a poet's privilege of originality in presentation—a feature that mostly shocks reviewers, who are too apt to pronounce any verse that is not a precise echo of older verse to be odd, eccentric, not poetry, etc.' (*Collected Letters*, iii. 279).

As he began his renewed poetic career, Hardy read again Poe's 'Philosophy of Composition'. Hardy had a high opinion of Poe. In 1901 he said that Poe's 'too small sheaf of verse has genius in every line, as well as music' (*Collected Letters*, ii. 303). In 1909 he said that Poe 'was the first to realize to the full the possibilities of the English language in rhyme and alliteration', especially in 'Ulalume' (*Life*, 343). But in 1901 Hardy found the 'Philosophy of Composition' 'pure fiction' (*Collected Letters*, ii. 303). Presumably Hardy could not believe the a priori way Poe claimed to plot all the effects of 'The Raven'. But Hardy may have been struck by one paragraph:

My first object (as usual) was originality. The extent to which this has been neglected, in versification, is one of the most unaccountable things in the world. Admitting that there is little possibility of variety in mere *rhythm*, it is still clear that the possible varieties of metre and stanza are absolutely infinite—and yet, *for centuries, no man, in verse, has ever done, or ever seemed to think of doing, an original thing.* The fact is, that originality (unless in minds of very unusual force) is by no means a matter, as some suppose, of impulse or intuition. In general, to be found, it must be elaborately sought, and although a positive merit of the highest class, demands in its attainment less of invention than negation. (Poe's italics.)

'Now, each of these lines, taken individually', Poe says, 'has been employed before, and what originality the "Raven" has, is in their *combination into stanza*; nothing even remotely approaching this combination has ever been attempted.' This passion for technique, we have seen, is behind Hardy's interest in verse skeletons.

Of course, this same passion is also the cause of critical misunderstanding of Hardy's metres—a point to which we will return.

Hardy's stanza forms—his rhyme schemes, his types of lines, his combinations of lines—provide an interesting study in themselves. Such a study reveals, in specific technical ways, the development of the poetry. Hardy combines and recombines the technical constituents of his verse throughout his career. As he does so, he makes new experiments with their multiple associations. Samuel Hynes writes in his influential study: 'One asks, "Why did he make that line shorter, and that line longer?" The answer, in Hardy's case, is usually that there *is* no apparent reason' (75). 'He evidently had little sense of the "emblematic" aspect of metre— the associations which have, through traditional usage, gathered around certain forms like the sonnet and the ballad.' The evidence is against this view. Yet so unknown is Hardy's metrical achievement that he is supposed to represent the metrical bankruptcy of the Victorian age rather than—what is the case—its metrical triumph. What we should do for Hardy is what De Selincourt said Saintsbury does for all of English metrics, namely study 'the growth of the constituent elements or limbs of the prosodic organism— of pause, rhyme, vowel-music, feet, metres, and so forth' (71). John Hollander has offered a justification of this kind of study:

To analyse the metre of a poem is not so much to scan it as to show with what other poems its less significant (linguistically speaking) formal elements associate it; to chart out its mode; to trace its family tree by appeal to those resemblances which connect it, in some ways with one, in some ways with another kind of poem that may, historically, precede or follow it. (*Vision*, 162.)

The Metrical Constituents of Hardy's Verse

In a number of places Hardy gives us hints toward defining the parts or constituents of his verse. In 'Rome: The Vatican: Sala delle Muse' (i. 136), published in 1901, he describes the aspects of poetry which fascinate him:

> 'To-day my soul clasps Form; but where is my troth
> Of yesternight with Tune: can one cleave to both?'
>
> .   .   .   .   .
>
> —'But my love goes further—to Story, and Dance, and Hymn,
> The lover of all in a sun-sweep is fool to whim—

The Muse assures Hardy that these are all phases of one inspiration. In 1908 Hardy arranged Barnes's *Select Poems* into tentative categories and acknowledged: 'many fine poems that have lyric moments are not entirely lyrical; many largely narrative poems are not entirely narrative; many personal reflections or meditations in verse hover across the frontiers of lyricism' (*Personal Writings*, 77). In 1912 Hardy agreed with a critic that such a division 'is frequently a question of the preponderance, not of the exclusive possession, of certain aesthetic elements'.[16] The implication to be drawn from these three statements is that Hardy thinks of a poem as composed of certain aesthetic elements associated with lyric and narrative, with form, tune, story, dance, and hymn. Combinations of these are the very stuff of Hardy's poetry.

What I would suggest is that we can tie such aesthetic elements to the metrical constituents which Hardy himself associated with dance and hymn, and so on. Association of certain metres with certain themes is, of course, an old story, as old as Plato; but in Hardy these constituents can be defined fairly simply, so that their pattern of development can be clearly seen. I like Hynes's notion of the emblematic aspect of a line and assume that a traditional line may carry with it

---

[16] *Life*, 359; the quotation comes from 'Modern Developements [*sic*] in Ballad Art', *Edinburgh Review*, 213 (1911), 153–79: see p. 153. Hardy's division of his own poems in *Selected Poems* (d. 1916) reflects this notion of preponderance: 'Poems Chiefly Lyrical', 'Poems Narrative and Reflective', etc. 'Preponderance' probably recalls Wordsworth's discussion of his classifications, based on 'the powers of mind *predominant*', in his preface of 1815. Hardy also knew Arnold's critique of these divisions (in 'Wordsworth'). One of the reviews of *Time's Laughingstocks* in Hardy's scrapbook stated: 'these subdivisions are quite arbitrary, since many of the lyrics are narrative, nearly all the narrative poems are in stanzas of lyrical quality, and love or the death of love is in all of them' (*Morning Post*, 9 December 1909).

associations which remain when the line is combined with other types of lines in a complex stanza.

The most prominent constituents of Hardy's verse seem to be the following:

1. *Iambic pentameter line.* In his notebooks Hardy copied from Theodore Watts's 'Poetry' in the *Encyclopaedia Britannica*, ninth edition: 'Before the poet begins to write he shd. ask himself which of these artistic methods is natural to him . . . the weighty (1) *iambic* movement, whose primary function is to state, or those lighter movements which we still call for want of more convenient words (2) *anapaestic* & *dactylic*—whose primary function is to suggest . . .' (*Literary Notebooks*, ii. 66). 'When Watts wrote to praise *Wessex Poems*, Hardy replied and noted his 'trepidation' at having his poems read by the author of this 'masterly essay'—*Collected Letters*, ii. 216.) We have seen that Hardy's apprenticeship as a poet began largely with the iambic pentameter, with some early blank verse, with an imitation of the Spenserian stanza, and above all with the sonnet. Hardy's thirty-eight sonnets tend to show a development from conventional imitation through modified conventions to original forms. The four regular Shakespearian sonnets (i. 19, 284, 285, 286) were written between 1865 and 1867. A Spenserian sonnet (i. 171) was written in 1867. Nine sonnets, written in 1866 or 1867 (one perhaps begun in 1865 or perhaps 1863), plus one other written in 1871, combine a Shakespearian octave with a Petrarchan type of sestet. The eleven Petrarchan sonnets are dated 1895 or later (with the possible exception of 'At a Lunar Eclipse'—see Chapter 2 n. 2 above). Some of the sonnets written later use rare Petrarchan sestets. Of the ten Wordsworthian sonnets, one was written in 1866, the others in the 1890s or later. The latest of these (ii. 331), dated 1921, uses a sestet, part of which rhymes with the octave in an apparently unique manner. Hardy's sonnet with the most unusual rhyme scheme (i. 332) was published in 1909. In general we can say that Hardy wrote all his Shakespearian sonnets between 1865 and 1867, combined Shakespearian octaves with Petrarchan sestets between 1865 and 1871, tried a lone Spenserian

experiment in 1867, and wrote a Wordsworthian sonnet in 1866. When he 'returned' to poetry in the 1890s, Hardy picked up with the Wordsworthian sonnet and also tried various kinds of Petrarchan sonnets—and used these two forms thereafter, sometimes in very unusual ways. (See Metrical Appendix, below.)

Throughout his career, we can see Hardy experimenting with the weighty iambic movement of the pentameter, and also trying combinations of that line with other kinds of lines. Hardy's skilful combination of these types of lines has gone largely unnoticed. Eventually Hardy tends to abandon the sonnet—at least two-thirds (27 out of his 38) of his sonnets are finished by 1901. He will continue, however, to use the pentameter and portions of the sonnet stanza in combination with other constituents.

2. *Iambic tetrameter and trimeter lines.* These two lines are associated in Hardy's most common traditional forms, the hymnal and ballad stanzas. Seventy-eight of Hardy's poems tend to follow 'short', 'long', or 'common' metre, while seventy-nine others tend to follow one of the other forms in Tate and Brady or in *Hymns Ancient and Modern.* Other common forms, which Hardy took from earlier poets, are too numerous to list here, but are listed in the Metrical Appendix. When tetrameters and trimeters are used in combination with other constituents, they tend to carry with them the ancient and popular echoes of hymn and ballad.[17]

3. *The dimeter line* in either rising or falling rhythm. This carries with it a strong lyrical resonance, due to its use as refrain in hymns and popular songs. It may have evolved out of the

---

[17] On the importance of the 1860s for hymn and ballad collections, see above. For a modern account of the relationship between long, common, and short metres, and their many variants, see Malof, chapter 4, 'Stress-Verse: The Native Meters'. The genesis of the ballad stanza and its relation to Latin and Old English sources is still a perplexed issue. On the relationship of English verse forms in general, see Schipper, about whom Brogan notes: 'It has been said that Schipper's table of contents is the most thorough, logical, and minute organization of English verse forms available anywhere; to peruse it is an education in itself. (Some have not thought so, however. . . .)'. On Schipper, see also my Metrical Appendix. An early synchronic account of verse structure is Puttenham's, ii. 10: 'Of Proportion by Situation'.

concluding short line of the sapphic stanza (see below, Metrical Appendix). Thirty-six of Hardy's poems are composed in dimeter lines alone. Thus Dylan Thomas speaks of 'the little tender short-lined simply rhymed and refrained love-lyrics' (*Evening*). In thirty-three of the fifty poems which Hardy referred to as 'songs', he uses the dimeter line. Dimeter lines are used in combination with other line types in 309 poems. When combined with other lines, the dimeter lines serve many functions—as refrains, as linking lines, as bob lines, as striking contrasts with longer lines. They often offer a sweet lyrical lilt which blends nicely with the traditional hymnal movement or the reflective pentameter line.

4. *The long line*, including alexandrines, hexameters, heptameters, octometers. Usually this type of line is used in combination with short lines; generally it seems highly experimental, carrying with it echoes of Renaissance experiments, and also reflecting what Saintsbury called the '*Long Metres* of Tennyson, Browning, Morris, and Swinburne' (*Historical Manual*, 115) (not to be confused with 'long' hymn metre cited above). These lines are often vital to some of Hardy's most interesting metrical effects. Used in combination with the shorter lines, they can create a strong expanding and contracting movement.

Most of Hardy's stanzas which use the long line are unique to him—with the exception of the common $a^7a^7b^7b^7$ form cited below (Metrical Appendix) and the interesting parallel of 'Beeny Cliff' (291) to a form ($a^7a^7a^7$) liked by Campion, which Hardy could have seen in his copy of *The Shakespeare Anthology*. Such long lines are also easily reducible to other constituents; indeed the fourteener easily breaks (as it did historically) into a combination of tetrameter and trimeter lines.

In Chapter 4, we can trace how Hardy combines and recombines constituents of his verse to create more complex mimetic effects.

Hardy was also very interested in the technical organization of language sounds. In the back of his copy of Walker's *Rhyming Dictionary*, he made some notes grouping letters and the sounds they represent: according to whether they are 'stopped' sounds (vowel followed by *b, c, d, g, k, p, q, t*) or 'open' sounds (*a, e, f, h, i, j, l, m, n, o, r, s, u, v, w, x, y, z*)); also according to whether they are 'kindred' sounds (*b = p, m, v; d = th, n, t; f = v, g = k, ng*, and so on). He groups consonants according to whether they are liquids (*l, m, n, r*), mutes (*p, f, b, v, t, th, d, k* [*ch*], *g*), sibilants (*s, sh, z, zh*), aspirants (*h*). He makes an alphabet which includes double consonants (*a, b, bl, br, c, cl, cr, d*, etc.). He also lists all the dipthongs associated with each of the vowels:

> h*e*—e, ee, ea, ei, eo, oe, ey, ie, y, i
>
> (e)
>
> *e*bb—e, ea, ie, eo, ue
>
> *i*l—i, y, ie, ye, ey, iu
>
> (i)
>
> *i*s—i, iu, ui, y   [etc.]

And he lists the combinations that go with consonants: i.e. '(k)—k. c. ch. g', '(m)—m. mn. lm. mb', and so on.

In this study, however, I have not attempted to trace Hardy's development of assonance, or his use of devices like those 'ingenious internal rhymes' and 'subtle juxtaposition of kindred lippings and vowel-sounds' which he found in Barnes (*Personal Writings*, 80). Hardy was sceptical about precise calculation of such effects. Citing Quiller-Couch's description of the sequence of vowels in Yeats's 'The Lake Isle of Innisfree', Hardy questioned 'whether any great poet ever thought much about such sequences'. Like the Victorian prosodists, Hardy was not ready to theorize about this dimension. An essay like David I. Masson's 'Word and Sound in Yeats' "Byzantium"' lay well in the future. Of course, Hardy's poems are full of such sequences, though I do not attempt to analyse them here. 'The Last Signal' is perhaps the most self-conscious example. The poem is a tribute to William Barnes which follows Barnes's instructions in his *Philological Grammar* concerning union (the rhyming of the end word of one line with the middle word of the next) and *cynghanedd* (consonant patterns within the line). But Hardy's most conscious artistic interest in sound symbolism was in his verse skeletons, and their symbolic or ironic possibilities. Here he was ahead of the Victorian prosodists.[18]

Hardy and the Theory of Dipody

There is another important way in which Hardy developed the insights of Victorian prosody. One of that prosody's most interesting insights, achieved by Patmore, concerned the phenomenon of dipodic rhythm. The idea of dipody helps answer an important theoretical question in Victorian prosody: how the metrical line can be capable of enormous, though not unlimited flexibility. The flexibility consists not only in variations of a metrical norm, but also in the same line's susceptibility to different metrical norms.

A good example of this flexibility is Hardy's 'After a Journey' (ii. 59), perhaps his most metrically interesting poem. F. R. Leavis has well described the sincerity of Hardy's voice in this poem, its cautious exploratory mode, its quiet enforcement of the paradox that 'to remember vividly is at the same time, inescapably, to embrace the utterness of loss' (133). But Leavis does not explain the metrical technique that contributes to the achievement. The first line of the poem would seem to be a pentameter, the second a tetrameter. Yet the balance of the poem is written clearly in a tetrameter rhythm,[19] and the uniformity of the indentation indicates here that the whole poem is written in a uniform rhythm. Other Hardy poems, 'The Sailor's Mother' (ii. 442), 'A Sound in the Night' (ii. 446), 'The Interloper' (ii. 230), begin in a similarly slowed tempo until the predominant tetrameter emerges. The first stanza may be scanned as follows, a scansion which reflects a recorded reading of the poem by Richard Burton:

---

[18] *Collected Letters*, v. 151. Hardy owned Quiller-Couch's *On the Art of Writing* (Cambridge, 1916, Cox catalogue). Another thing this book does not do is analyse Hardy's rhyme schemes. Again, what we need is someone to do for Hardy what Marjorie Perloff did for Yeats in *Rhyme and Meaning in the Poetry of Yeats*.

[19] Davie had a good intuition when he suspected that 'After a Journey' may be 'four-foot trochaic-dactylic' but then he strangely decides for 'English hendecasyllables' (57).

Héreto I cóme to view a vóiceless ghóst;
  Whíther, o whíther will its whím now dráw me?
Úp the cliff, dówn, till I'm lónely, lóst,
  And the únseen wáters' ejáculations áwe me.
Whére you will néxt be thére's no knówing,
  Fácing round abóut me éverywhére,
    With your nút-coloured háir,
And gráy eyes, and róse-flush cóming and góing.

Ian Holm's recorded reading confirms Burton's, with the interesting difference that Holm reads:

    Héreto I cóme to view a vóiceless ghóst

Also Holm reads line 5 with more or less level stress; indeed Burton's reading is barely differentiated. To these differences we shall return.

The remaining lines of the poem are also read by Holm and Burton in a tetrameter manner; the 'feet' vary from initial single syllable truncated feet to tetrasyllabic feet as in the following:

  Whí | thĕr, ŏ whí | thĕr wĭll ĭts whím | nŏw dráw (mĕ)

There are no internal monosyllabic feet, thus distinguishing this traditional accentual-syllabic rhythm from Hopkins's sprung rhythm.

The issue of tetrasyllabic feet was of great theoretical interest for the Victorians because it bore on the question of how much variety can be maintained within a given metre. How many unstressed syllables may fill one position, or how many syllables may fill the bar? De Selincourt cites the extreme equivalence in Tennyson's 'Break, Break, Break' (where 'Break, break, break' and 'On thy cold gray stones, O Sea' are equivalent lines) and asks: 'If this can happen why, we may . . . ask, should not feet of four or five syllables be allowed?' (83) Indeed the ballad 'Lord Randal' was well known; and recent examples like Poe's 'Ulalume' ('The léaves they were withering and sére') and poems by Swinburne were often cited by prosodists as examples of poems using tetrasyllabic feet (Omond, *Study*, 96 ff.). Hardy did not hesitate to revise the

dimeter last line of 'The Souls of the Slain' from 'Sea-moanings
and me' to 'Sea-mutterings and me'.

Many examples of folk songs, with strings of three unstressed
syllables, yet organized into equivalent spaces under the
influence of the musical bar, can be found in Peter Kennedy's
Folksongs of Britain and Ireland (New York, 1975). In 'Green
Grass It Grows Bonny' (357) the line, 'I wonder what is
keeping my true-love tonight', resembles Hardy's line, 'I see
what you are doing: you are leading me on'. In 'The
Nobleman's Wedding' (364), many syncopated lines, like
'Early next morning when her husband awakened', resemble
Hardy's syncopated lines, like 'Whither, O whither will its
whim now draw me'. Hardy knew a version of this last song as
'The Unconstant Lover'. Other examples Hardy knew of folk
songs with such strings of unstressed syllables were 'The
Beggar's Wedding' (Udal, 308), ' "It's hame, and it's hame" '
(*Mayor of Casterbridge*, 8), and various examples in Hardy's
1799 Music Book. In 'How happy's the man', for example, the
first line, 'How happy is the man that free from care', must be
adjusted to the emerging tetrameter beat.

Saintsbury for one was uncomfortable with the notion of
four-syllable feet and believed that they could be resolved into
combinations of shorter feet. If the paeon was resolved, as
Saintsbury wished, into smaller feet, the result in the above
example would be a pentameter line. 'How happy is the man
that free from care', in Hardy's 1799 Music Book, becomes,
with Saintsbury, 'How happy is the man that free from care'.
'After a Journey' is also susceptible to similar *Gestalt* switches.
Indeed, many of the lines, like the first, can be read as
pentameters. Within the paeon, Hardy shows, there nestles just
such an alternative possibility.

Interestingly, the debate between pentameter and tetrameter readers of 'After a Journey' resembles strikingly the
debate between Saintsbury and De Selincourt over the question of the metre of Spenser's *The Shepheardes Calender*. Saintsbury saw in 'February', 'May', and 'September' a tendency for
the anapaestic tetrameter lines 'to slip into actual decasyllables,
more or less normal' (i. 355). In spite of this, he believed that

the 'base' of 'February' 'is a four-foot (or "four-accent") line,
which is capable of being reduced without injury to its norm of
eight syllables, and of being extended, also without injury, to
twelve, anapaests being by equivalence substituted for iambs'
(i. 353). Saintsbury says that the verse becomes more obviously
anapaestic in 'May', and even more so ('therefore less interesting') in 'September'. Like 'After a Journey', 'May' begins with
a pentameter-like line: 'Is not thilke the mery moneth of
May'.

De Selincourt agreed with Saintsbury's tetrasyllabic reading,
but denied the 'intrusion of five foot lines'. Many of the lines, he
says, 'make excellent decasyllables, if they are taken out of their
context; but to read them so in the poem is to break the lilt and
change the tone of it completely. How much too the four-
syllable feet we indicate add . . . not only of emphasis but also
of colloquial charm! Instead of an unconvincing blend of
disparate metres, the poems exhibit, surely, a tripping, stum-
bling use of the native style, the virtues of which flow in the
blood of all good Englishmen' (88–9). 'Feet of four syllables,
which Prof. Saintsbury disallows, are surely quite common'
(94).

De Selincourt's solution—'a rough rhythm grouped about
four accents' (88–9)—is a little too easy. It does not do justice to
the sense of a measured accentual-syllabic rhythm we get from
some pieces in *The Shepheardes Calender*, a sense which drove
Saintsbury to classify various lines as tetrameters or pentameters 'intermingled at random', as De Selincourt put it
(88).

The Shepheardes Calender is a key work in the history of English
metrical rhythm and has been a puzzle for centuries; 'After a
Journey' returns us to its issue, an issue at the very heart of all
our metres. There has been a persistent school of thought which
sees an older four-beat 'native' rhythm that continues to be
heard in our pentameter poems.[20] Francis Gummere's article of

---

[20] Frye is a prominent modern spokesman: see *Sound the Poetry* (New York, 1957),
xvii, and *Anatomy of Criticism* (Princeton, 1957), 251 ff. In a letter to me, Brogan cites
Cobb's articles (dated 1910–17), Malof, and Frye (Brogan, *English Versification*, 197–8,
361).

1886 is the classic statement in English of this position. Drawing
on Schipper and Ten Brink, Gummere saw Chaucer's 'wonderfully flexible verse' as 'a harmonizing . . . of the two great
systems, the Germanic and the Romance, the rhythmic and the
metric, *on the basis of two representative measures*' (54–5). 'On the
one hand *four* stresses, fixed pause, indeterminate amount of
light syllables; on the other, *five* stresses, shifting and slighter
pause, strict ordering and number of light syllables.' (57) In
some detail Gummere uses *The Shepheardes Calender* especially
'February', 'May', and 'September' to show how 'our heroic
verse . . . is *simply the result of forcing the iambic movement (influence
of foreign models played its part here) upon some late form of our old
four-stress verse*' (62, Gummere's italics). Gummere saw this
tendency toward the heroic line within Spenser's tetrameter
lines: '*Nearly 10 per cent. of the verses will allow the iambic movement,
and so become heroic verse*' (64, Gummere's italics), as for example
line 89 from 'February', 'Whose way is wilder<u>nesse,</u> whose ynne
Penaunce' (where I have underlined the potential fifth accented syllable Gummere refers to). Conversely, Gummere often
saw in the standard heroic line a weak accent capable of being
submerged back into the tetrameter, as in 'No secret island *in*
the boundless main' (58, Gummere's italics). This was North-
rop Frye's point many years later.

   The Shepheardes Calender is a *locus classicus* in the relationship of
these basic forms of our verse, the rough four-beat native verse,
the more measured tetrameter, and the measured pentameter.
My·sense of 'February', 'May', and 'September' is that Spenser
is showing how the seemingly rough rhythm felt by De
Selincourt can in fact be ordered into a measured tetrameter.
We are able to redistribute the stresses in a way that gives us a
conceivable though at first unapparent pattern of speech
stresses to fit the metre. (Here are some possible, but not
inevitable, ways in which the metre can be maintained.)

Must nót the world wénd in his cómmun cóurse   ('February', 11)
Bút such éeking hath máde my hart sóre   ('September', 31)
For Pán himselfe wás their inheritánce   ('May', 111)

The lines are therefore theoretically resolvable.[21] As Saintsbury suggested, *The Shepheardes Calender* may be the first work in English in which a poet '*deliberately* adopted the process of Substitution'.[22] I think this means that some sections of *The Shepheardes Calender* self-consciously dramatized how different distributions of syllables could 'fit' feet, and so 'save' the metre. There is much good humour in the experiment.[23] I do not find the same complexity in Spenser's pentameter months. Perhaps Spenser saw an evolution from strong stress verse to common metre, which then prepared the way for the mature pentameter where the romance and Saxon elements are perfectly blended.

It is appropriate that Hardy's poetic career more or less began with this poet's poet when he attempted to translate *Ecclesiastes* into Spenserian stanzas (*Life*, 47). 'After a Journey' seems to reflect the issue that is intriguing Spenser at the beginning of the modern metrical tradition. The poem seems to recapitulate the broad historical development of strong stress verse into tetrameter verse *and* into pentameter verse, and all implicit in a single line. The four-beat native rhythm seems to shape into equivalent intervals, and within this equivalenced line nests the possibility of the pentameter.

The riddle therefore remains: how can the two systems, pentameter and tetrameter, work together so gracefully in this poem. As Gummere said, 'Of course, I do not undertake to say *how* Chaucer . . . combined the two elements' (56). The answer, for 'After a Journey', may lie in the article with which the new prosody began: Patmore's essay of 1857. Patmore, we may recall, was one of those who said that poetry represents an intensification of the rhythm of prose. But Patmore added an interesting addition to this traditional theme. In a passage Hardy copied, Patmore stated 'the great general law, which I

[21] Note that my scansions give more of a measured lilt than John Thompson's scansions, 98–9.

[22] Saintsbury, i. 382. About this claim, De Selincourt commented: 'If Prof. Saintsbury's account is the true one, his system is first consciously used in poems whose rhythm is so informal that the critic remains in doubt whether the base of the lines is iambic or anapestic. . . .' (88).

[23] Redistribution of stresses in Wyatt, by contrast with Spenser, seems often to require an impossible pattern of speech stresses. Thus Wyatt remains a metrical riddle.

believe that I am now, for the first time, stating, that the *elementary measure, or integer, of English verse is double the measure of ordinary prose,*—that is to say, it is the space which is bounded by *alternate* accents; *that every verse proper contains two, three, or four of these 'metres,' or, as with a little allowance they may be called,* "*dipodes*" ' (25–6, Patmore's italics). 'All English verses', Patmore went on (and Hardy copied), 'in common cadence are therefore dimeters, trimeters, or tetrameters, and consist, when they are *full*, i.e., without *catalexis*, of eight, twelve, or sixteen syllables' (26). Patmore noted Shakespeare's tendency toward dipody in the following iambic pentameter lines which he called 'blank trimeters':

> The crow doth sing as sweetly as the lark
> When neither is attended; and I think
> The nightingale, if she should sing by day,
> When every goose is cackling, would be thought
> No better a musician than the wren.[24]

Citing the 'unusually distinct and emphatic accentuation of the first syllable in the metrical section'. Patmore said that in 'these blank trimeters, properly read, there is a major and a minor accent in every section but one'. When 'full', therefore, such lines would consist of twelve syllables, the final pause after each line counting for two syllables. A quarter of a century later, Patmore was still excited by this insight. In 1881 he said in a letter to a friend that he had intended to give 'further proofs of the essentially *dipodal* character of all metre'.[25]

---

[24] Patmore, *Essay*, 29. I have used double slash marks (where Patmore uses single ones) to indicate the dipodic primary stress. Patmore leaves out the dipodic accent on 'thought' in the 1857 version, presumably the 'section' he excludes, perhaps because the line is run-on. He also leaves out the dipodic accent on 'every' in subsequent editions, perhaps because he excludes adjectives from primary category words here (see Longmuir, above). Sister Roth speculates that Patmore invented the spelling 'dipode', but in fact he got it from John Mason (see below). The *OED* missed this spelling and finds the earliest use of the word 'dipody' in 1844. The *OED* finds the word 'dipode' used only in reference to bipeds. In an 1860–1 *Philological Society Transactions* essay, Thomas Barham asserted that it is 'more convenient, as well as more agreeable to ancient usage, to take the rhythm *by dipodies*' (51).

George Stewart's 1924 article on 'Dipodic Verse' missed Patmore and gave Lanier credit for the 'first recognition of dipodic structure' (979).

[25] *Memoirs*, ii. 268. Patmore continues in the *Essay*: '*The so-called 'Alexandrine', at the*

Hopkins was particularly impressed by Patmore's notion of 'dipodes' (a spelling Patmore apparently took from John Mason). On 26 January 1881 Hopkins discussed the *Essay*: 'The principle, whether necessary nor not, which is at the bottom of both musical and metrical time is that everything shd. go by twos and, where you want to be very strict and effective, even by fours . . . the instance Pat gives is good and bears him out' (*Letters*, 119). Patmore discussed the 'alex-andrine' of Drayton's *Polyolbion* ('a tetrameter, having a middle and a final pause each equal to a foot' (44)) in a way that accords with Hopkins's understanding of a similar principle in *St. Winefred's Well*' (*Letters*, 212). Hardy marked several passages in the first three 'Songs' of Drayton's *Polyolbion*, and cited it twice in *Jude the Obscure* (IV. i. 239, iv. 275), his last great novel, after which he turned to poetry. In a letter of 1904 Hardy said: 'I know Drayton's Polyolbion better than the rest of his work' (*Collected Letters*, iii. 133). It would seem that both Hardy and Hopkins, both nourished by the 'new prosody', were influenced by the phenomenon of dipodic metres at an important stage of their poetic careers.

Dipodic metre, traditionally considered a rare and rather perplexing phenomenon, reveals an important fact about English metrical rhythm. Patmore's claim that the measure of verse is double the measure of prose was mysterious and certainly controversial to some of his readers. One sense we can make of it is that verse reveals the structural principle of prose rhythm. Prose organizes its stresses into patterns which are subsets of larger patterns, all governed by alternating stress principles. The dipodic phenomenon of double iambs (Fig. 1) is a small mirror of the hierarchies in prose stresses.[26] The dipody exaggerates the hierarchical principle and so seems (to Patmore at least) to double the speed of the prose rhythm.

---

end of the Spenserian stanza, is quite a different verse, though including the same number of syllables; it is the mere filling up of the trimeter; and that Spenser intended it so is proved by the innumerable instances in which he has made middle pause impossible' (44).

[26] Kiparsky, 229, points out: 'The dipodic character of many English meters can be explicated by assuming that feet themselves are bracketed into larger units.' Also see Attridge, *Rhythms*, 116.

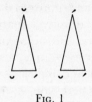

FIG. 1

Traditionally, of course, many Latin and Greek metres were organized into groups of two feet, so that a Latin trimeter for example was in fact three double feet, dividing the six feet we would now usually call a hexameter. John Mason in 1749 applied the various kinds of classical '*Dipodes* or double feet' to English polysyllables ('alabaster' is a 'third Paeon'), but concluded that such feet are 'of no Use, expecially in English Metre' (*Essay on . . . Poetical Composition* 17, 25). Patmore was the first to realize the potential interest of the issue: 'It is difficult to discover how far this general law of English verse has been felt by prosodians. Certainly it never has been fairly expressed by them, though Foster gives the English heroic line the name of its Greek counterpart, whereby he assumes such division' (25). That the same line can be described as a trimeter or a hexameter, that a line consists of multiples made out of multiples, is the lesson dipodic metres teach. For example, the line quoted earlier from Hopkins,

A pénniless advénturer is óften in extrémities

can be read as a tetrameter line or as a heptameter (eight-accent) line if those blank stresses are 'realized'. Dipodic metres show that it is possible to realize a number of different patterns in the same line. In fact, more than two patterns can be realized—depending on how many accents are actualized. The dipodic phenomenon has a special relation to blank accents, with their implicit power of being actualized.

The dipodic phenomenon also represents the frontier of the Victorian understanding of metrical form. Omond, rejecting Patmore's notion of final feet to fill out the dipody, did not further consider the issue. In his *Study of Metre* (105), he puzzled over the phenomenon of 'quadruple measures' which 'can always in the last analysis be resolved into duple time'. Saintsbury believed the same, though he acknowledged that 'occasional syzygies [i.e. dipodies], however, as a kind of extra equivalence, *may* suggest themselves not infrequently to some ears. . . . But the subject is a really interesting one'; and he looked to 'future references to the subject of foot-composition and distribution' (iii. 137–8). Thus, I am suggesting that 'After a Journey' is a climactic moment in Victorian prosodic theory even though the theory had yet to be stated fully.

Bridges, however, with his superb ear more open to freer forms of accentual-syllabic rhythms because he had divorced them from metrical counting (see above, Chapter 1), described the phenomenon in his 1901 edition of *Milton's Prosody* (104). Shelley's 'Away, the moor is dark beneath the Moon' uses a 'skeleton scheme of four double stresses and the break in the middle'. Thus the lines can vary from ten to fifteen syllables. (Notice that the first line could at first be read as iambic pentameter)

> Awặy! the mŏor is dặrk beneath the mŏon
>   Rắpid clouds have drŭnk the lắst pale beam of ĕven.
> Awặy! the gathering wĭnds will cắll the darkness sŏon,
>   And profŏundest midnight shrŏud the serĕne lights of hĕaven.

Responding to Henry Bradley's praise of this explanation, Bridges wrote: 'I was most pleased myself with the explanation of . . . one of the most beautiful and affecting poems . . . excluded from collections because the pedants (like Palgrave) thought that it didn't scan' (*Selected Letters*, i. 402). Bridges notes the same rhythm in a song in *Prometheus Unbound*, i. 772–9. Dipody makes clearly evident the power of the metre to exalt some syllables and depress others for its rhythmic purposes. Later, George Stewart would exhaustively discuss the ancestry and use of dipody in Romantic and later verse.[27]

---

[27] In 1924, Stewart, in 'A Method Toward the Study of Dipodic Verse', suggested a simple way of measuring the dipodic effect—by counting how often normally weak stresses appear in the place of the alternate accents. On dipodic metres, also see his *Modern Metrical Technique*, and Malof, 126 ff.

Interesting theoretical discussion of dipody then had to wait until Wimsatt and Beardsley's classic restatement of the new prosody (though they were forgetful of the Victorians) in 'The Concept of Meter: An Exercise in Abstraction'. The pheno-menon of dipody nevertheless undermines Wimsatt and Beardsley's definition of accentual-syllabic metres as depending simply on 'an approximately equal number of weaker syllables between the strong stresses' (115). The dipodic possibility indeed allows a four-beat line to consist theoretically of anywhere from four to sixteen syllables. So Wimsatt and Beardsley found the phenomenon perplexing and tried to dismiss it.

The notion of dipodic structure, then, was so far in advance of its time that later Victorian theorists (except Hopkins) rejected the idea and modern theorists (like George Stewart and Wimsatt and Beardsley) ignored Patmore's essay. Even Hardy, when he first read the essay, said that some of its ideas were 'suggestive—others doubtful' (*Literary Notebooks*, ii. 193). However, by the time he wrote 'After a Journey', half a century after Patmore's essay, he was ready in poetic terms to come to grips with Patmore's insight. We know that he took notes from Patmore's essay about six years before he wrote 'After a Journey'. As we shall see, most of Wimsatt and Beardsley's points are illustrated in 'After a Journey': the popular narrative base of the metre, the 'strong-accent meter . . . with number of syllables and minor stresses tightened up into a secondary pattern', the perplexed 'choice just where to place the dipody', the 'first feeling . . . that a strong lilt or swing is present though it is hard to say just how it ought to be defined', the great variety in number of syllables per line, and the way in which the falling metre may convert to rising metre. We also know that Hardy was influenced by several dipodic poems.

Scott's 'The Eve of St. John' is a widely noted early example of such intriguing metres; and it is significant that Hardy said this poem was 'among the verse he liked better than any of Scott's prose' (*Life*, 239; also see *Collected Letters*, v. 174). The poem also intrigued Saintsbury who believed it influenced later examples of lines with extra-metrical syllables ('There is considerable prosodic interest in "The Eve of St. John" ', iii. 81 n. 1).

> In slĕep the lády mŏurn'ed, ánd the Bãron tóss'd and tŭrn'd
>   And oft to himself he said:—
> 'The wŏrms aróund him crĕep, ánd his blŏody gráve is dĕep
>   It cannot give up the dead'.

The ballad is a typical tale of revenge and a *revenant* ghost— here the ghost of a love, killed by the baron 'near Tweed's fair strand' and conjured up by the baron's guilty unhappy wife: 'At our trysting-place, for a certain space, / I must wander to and fro'. That Scott's story connects with the story of 'After a Journey' is a measure of how far Hardy has come in adapting the Gothic ballads of his younger years to the lyric anguish of his maturity.

It is interesting that Meredith's 'Love in the Valley'[28] is also similar in theme to 'After a Journey' and describes a meeting with a well-beloved 'Borne to me by dreams when dawn is at my eyelids; / Fair as in the flesh she swims to me on tears':

> Hĭther she cŏmes; she cŏmes to me; she lĭngers,
>   Dĕepens her brown ĕyebrows, whĭle in new surprĭse
> Hĭgh rise the lãshes in wŏnder of a strãnger;
>   Yĕt am I the lĭght and lĭving of her ĕyes.

Hardy will manage this dipodic structure in a much more gentle and subtle way as he meets his well beloved 'Facing round about me everywhere, / With . . . gray eyes, and rose-flush coming and going'.

'Love in the Valley' was written in 1851 and revised in 1878. Meanwhile in the 1860s, Tennyson had published 'In the Valley of the Cauteretz'. I quote it because of the striking likeness of theme and imagery to 'After a Journey':

> Ãll along the vãlley, strĕam that flashest whĭte,
> Dĕepening thy vŏice with the dĕepening of the nĭght,
> Ãll along the vãlley, whĕre thy waters flŏw,
> I wãlked with one I lŏved two and thĭrty years agŏ.

[28] De Selincourt, in his study of Meredith, sensed the dipodic structure of the poem, but could only describe it in terms of conventional substitutions: 'you are liable at any point in the line to have the trochee replaced by a single long accented syllable' (237).

> Ăll along the vălley, whĭle I walked to-dăy,
> The twŏ and thirty yĕars were a mĭst that rolls awăy;
> For ăll along the vălley, dŏwn thy rocky bĕd,
> Thy lĭving voice to mĕ was as the vŏice of the dĕad,
> And ăll along the vălley, by rŏck and cave and trĕe,
> The vŏice of the dĕad was a lĭving voice to mĕ.

It is interesting that the connection between these themes and the dipodic lilt should have come to fruition in Hardy's mind forty years later.

Dipody is also skilfully used in two of Browning's 'musical' poems which influenced Hardy. 'A Toccata of Galuppi's' (1855) gives us 15-syllable trochaic lines which are divisible into dipodic multiples.

Ĭn you cŏme with yŏur cold mŭsic tĭll I crĕep through ĕvery nĕrve

'Abt Vogler' (1864) is more subtle because it organizes the dipodic beat into codas, a little like Latin elegiac pentameters:

> Wŏuld that the strŭcture brăve, ‖ the mănifold mŭsic I bŭild,
> Bĭdding my ŏrgan obĕy, ‖ călling its kĕys to their wŏrk

When Hardy read the 1909 *Spectator* review of Swinburne's 'science' of metrics, he copied: 'Browning never approached Mr. Swinburne in the matter of metrical science. . . . [yet Swinburne] could not have written the "Toccata of Galuppi's"'.

Finally, 'After a Journey' also reflects the era of Kipling, Masefield, and other late Victorian and Edwardian poets, whom Hardy read and who experimented with pronounced dipodic metres.

This dipodic divisibility is more pervasive in English poetry than Wimsatt and Beardsley allow. Its possibility suggests that a sequence of syllables in a given line is part of a larger hidden structure which generates its rhythm. The basic rhythmic principle, binary, hierarchical, generates the metrical rule, i.e. tetrameter, pentameter, which can be changed without violating the principle; and the rule can generate a norm, actualizing whatever pattern of accents and non-accents the poet pleases, consistent with the possibilities of the language. In their dispute with Frye, Wimsatt and Beardsley were right to insist on their norm (iambic pentameter), but Frye was correct in his intuition of how easily a pentameter can metamorphose into a tetrameter. Some of Patmore's most subtle analysis comes in distinguishing lines that are the same superficially. He analyses the alexandrine in *Polyolbion* as a dipodic tetrameter as we have seen, and the alexandrine in the Spenserian stanza as a dipodic trimeter like the pentameters in the same stanza, with the alexandrine showing the full form (44). Both Hardy and Hopkins were fascinated with the different ways multiples can build out of multiples, like Abt Vogler's notes which 'dispart now and now combine', 'Raising my rampired walls of gold as transparent as glass'.

The possibility of dipody is common in Hardy's poetry.[29] Examples can be clearly seen in poems which we cited above in connection with 'After a Journey'. Thus, 'The Sailor's Mother' (ii. 442):

That brŏught you blíndly knŏcking ín this mĭddle-wátch so drĕar?

In 'A Sound in the Night' (ii. 446), the dipody seems to resolve the opening iambic-anapaestic ambiguity of the poem:

It seĕms to bé a wŏman's vóice; each lĭttle whíle I hĕar it

Dipodic rhythm seems to emerge quite sharply in 'Poems of 1912–13' in which 'After a Journey' is included. Thus 'Beeny Cliff' (ii. 62):

O the ŏpal ánd the săpphire óf that wăndering wéstern sĕa

Also 'His Visitor' (ii. 57):

I cŏme acróss from Mĕllstock whíle the mŏon wástes wĕaker

Hickson scans this line as pentameter and the other long lines in the poem, hopelessly, as hexameter, pentameter, and heptameter. J. B. Mayor made the same sort of mistake with Shelley's 'Away! the Moor is dark beneath the moon', whose true principle Bridges had discovered (Mayor, 1901 edition,

[29] In 1952, Ransom noted that Hardy 'made stanzas out of variations upon the "folk line", which nowadays we call the dipodic line', 'Hardy—Old Poet', 16.

250–2). Alfred Gordon made a similar mistake in the essay 'The New Prosody: A Rejoinder', which he sent to Hardy. There he scanned 'Love in the Valley' as a mixture of pentameter, hexameter, and heptameter lines.

In 'In Tenebris II' (i. 207) and perhaps 'Wessex Heights' (ii. 25), which is a little like 'Abt Vogler', the seven-beat lines seem to invite division into a larger multiple. 'Aquae Sulis' (ii. 90) starts with a steady iambic pentameter movement like 'After a Journey',

> The chimes called midnight, just at interlune,

which moves quickly into an anapaestic pentameter,

> And the daytime parle on the Roman investigations

which then seems subject to a dipodic division:

> Those of the Goddess whose shrine was beneath the pile

In this poem the strict iambic pentameter movement seems associated with stern Christianity ('Had set up crucifix and candle here') while the emerging anapaestic movement seems associated with the emergence of the old pagan goddess's ghost:

> And the flutter of a filmy shape unsepulchred,
> That collected itself . . .

The dipodic possibility seems to stand for the potential interchangeability of these two metrical norms and may even reflect the poem's theme: the interchangeability of Christian God and pagan goddess. 'After a Journey' encourages a somewhat similar interplay of present consciousness associated with the pentameter sound ('Hereto I come . . .') and of a ghost-invaded consciousness in the tetrameter sound ('Whither, o whither will its whim now draw me').

The mathematics of dipodic rhythms are interesting. Apparently, a rising and a falling rhythm, an even- and odd-numbered stress line, can nestle within the same dipodic line once we conceive the dipodic principle. This principle explains how the two rhythms, 4- and 8-beat falling, 3- and 7-beat rising, can 'ghost' each other in 'The Revisitation' (i. 238):

Through the gateway I betook me
Down the High Street and beyond the lamps, across the battered
    bridge

The possibility of two metrical readings is functional to the
poem's theme: they capture the contradictory directions of the
speaker's journey, advancing into the country, regressing into
the past.

The dipodic principle explains how so many different
variations of metrical norm can 'fit' a given metre. Though a
poet normally establishes a given metrical norm, Hardy is
intrigued by the principle on which all accentual-syllabic metre
is based, a principle which can emerge and call attention to
itself, not as an oddity of our metres, but as their basis. Such
elasticity of our language to serve a variety of tiers of metrical
structure explains how a line can have an extra foot and not
seem unrhythmical. Thus 'Where the Picnic Was' (ii. 69) is a
predominantly dimeter poem, but has the line: '—But two have
wandered far'. The pattern of stressed and unstressed syllables

u s   u s,

is merely reassembled (Fig. 2).

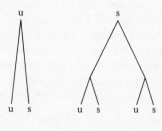

Fig. 2

The norm changes, but the dimeter spacing is preserved by the
hierarchy of stresses.

The dipodic principle explains how the two metrical norms
in 'After a Journey'—the deliberative reflective pentameter
and the lyrical rolling tetrameter—can be consistent with a

common alternating rhythm. Thus 'After a Journey' never
settles completely into a song lilt. In Gosse's *Gray* (London,
1882)—which Hardy told Gosse 'afforded me food for thought
during several days'—Hardy had read Gosse's distinction
between the simple 'flute-melodies' of Collins and the poetry of
Gray: 'In Gray the song, important as it was, seemed merely
one phase of a deep and consistent character, of a brain almost
universally accomplished, of a man, in short, and not of a mere
musical instrument' (65). This poise of song rhythm and
reflective pace is an effect we often find in Ben Jonson, whom
Hardy cited when he confessed an ambition to be in the *Golden
Treasury* (*Life*, 444). Another way of describing the balance in
'After a Journey' is that it brings together the tetrameter of
popular song stanzas and the pentameter of the Keatsian ode
which is similar in shape (especially 'Ode to a Nightingale').[30]

In the following possible scansion of the first stanza of 'After
a Journey', I have inserted blank accents which are the points
at which a different norm might be realized: use of these
accents reveals how the first line might break down into a
tetrameter norm, and how the others might build up into a
pentameter norm:

> Hereto I come to view a voiceless ghost;
>   Whither, o whither will its whim now draw me?
> Up the cliff, down, till I'm lonely, lost,
>   And the unseen waters' ejaculations awe me.
> Where you will next be there's no knowing,
>   Facing round about me everywhere,
>   With your nut-coloured hair,
> And gray eyes, and rose-flush coming and going.

---

[30] *Collected Letters*, i. 110; Saintsbury, ii. 155. Yvor Winters's discussion of Jonson's
'Though Beauty Be the Mark of Paradise' is very appropriate for Hardy: 'It is a fusion
of two kinds of poetry: the song and the didactic poem. . . . The stanza frequently
suggests a song stanza as it opens, and then seems to stop the song with a didactic close,
as if strings had been plucked and then muted; and this effect, more or less inherent in
the form of the stanza itself, is sometimes stressed and sometimes softened' ('Poetic
Styles, Old and New', in Don Cameron Allen (ed.), *A Celebration of Poets* (Baltimore,
1959), 63–4). In Hardy this effect seems to work in reverse. Irving Howe has some
perceptive remarks on this aspect of Hardy's metre (880–1, 905); but his article in
general repeats the traditional critiques of Hardy's mechanical verse forms, ill-assorted
diction, and fixated opinions.

In one of his revisions to the poem, Hardy illustrates his recognition of the dipodic principle. In his first version, the fourth line read:

And the unseen waters' soliloquies awe me

The anapaestic tetrameter is clear. The change of 'soliloquies' to 'ejaculations' in the revised version provides an extra dipodic syllable which theoretically supports the pentameter possibility. Hardy's first version of the first line of the last poem made the dipody clearer:

Héreto I cóme to ínterview a ghóst

Hardy may have changed this version for expressive reasons (see Leavis on 'view', 131) or he may have wanted to set up a strong sounding pentameter and follow it with a resounding tetrameter, thus announcing the metrical conflict. Indeed the two lines together illustrate Hardy's competing tasks: to make the fifth metrical stress strong enough to suggest the pentameter, but also to make it weak enough to be submerged by the tetrameter.

There is a further distinction to be made about dipody. In a sense, Wimsatt and Beardsley are right: explicitly dipodic poems are relatively rare, and Patmore goes too far in hearing dipody everywhere. Thus, Hopkins qualified Patmore's discussion: 'But whereas this ['the principle ... that everything shd. go by twos'] is insisted on and recognised in modern music it is neither in verse. It exists though and the instance Pat gives is good and bears him out' (*Letters*, 26 January 1881, 199, citing Patmore's *Essay*, 27). What *is* everywhere is the dipodic potential, the potential for stresses to be organized into multiples of stresses, and for these multiples to allow the actualization of other metrical rules and norms. From a self-consciously dipodic poem like 'Love in the Valley' to a discerned dipodic rhythm in standard lines like the Shakespeare lines dissected by Patmore to a standard line like Pope's heroic line (which Patmore complains avoids dipody (*Essay*, 43)), we see the continuity of English verse rhythm. The genius of

'After a Journey' is that it takes us up and down the scale of this history.

Hardy had been given a copy of *The Golden Treasury* in 1862. He wrote 'After a Journey' in 1913. The poem is not only a return to the olden haunts of Hardy's personal past: 'Through the years, through the dead scenes I have tracked you'. It is also a return to old metrical haunts. In its skilful ambiguity, it represents most of the major constituents of Hardy's verse. It could only have been written by a poet who had explored the multiple possibilities of these various lines and rhythms, and related their long history to his emotional life. To the expressive function of the poem's dipody, we shall return at the end of Chapter 4.

3. Hardy and the Tradition of Sound Symbolism

GIVEN their technical breakthrough in describing the nature of accentual-syllabic rhythm, the next step for the Victorian prosodists was to describe the aesthetic uses to which it was put. The last thing that Hardy summarized from Patmore was from his discussion of blank verse:

The great diff[iculty], as well as delight, of this measure is not in variety of pause, tone, & stress for its own sake. Such variety must be incessantly inspired by, and expressive of, ever-varying emotion. Every . . . deviation from the strict and dull iambic rhythm must be either sense or nonsense. *Such change is as real a mode of expressing emotion as words themselves are of expressing thought*: and where . . . [such an end does not exist] . . . the variety obtained is more offensive . . . than the dullness avoided. (*Literary Notebooks*, ii. 192–3.)

Earlier in his review of *In Memoriam* (*North British Review*, 13 (1850), 532–3), Patmore had applied this principle to all accentual-syllabic verse: 'Mere variety . . . ought never to be, or, at all events, to appear to be, the first motive in the introduction of irregular feet' (534). This, of course, was an ancient principle in prosody. What we see is that the Victorian understanding of the nature of metre made possible a richer understanding of the function of metre.

In studies of Hardy's verse, the accentual-syllabic principle of the verse is sometimes ignored. And the purpose of variation within the pattern is sometimes not understood. Thus, in answering his reviewers in his *Later Years* assessment, Hardy made two complaints whose full significance may not at first be apparent.

One case of the kind, in which the poem 'On Sturminster Foot-Bridge' was quoted with the remark that one could make as good music as that out of a milk-cart, betrayed the reviewer's ignorance of any perception that the metre was intended to be onomatopoeic, plainly as it was shown; and another in the same tone disclosed that the reviewer had tried to scan the author's sapphics as heroics. (*Life*, 301)

Both the identity of Hardy's verse patterns, as well as the 'onomatopoeic' purposes which they serve, must be understood. Thus, it is not enough to say that Hardy roughened his metres in the name of spontaneity. Intentionally 'inexact . . . rhythms' means an art which experiments with the complex relation between metrical form and speech rhythms in order to achieve certain purposes.

The History of Onomatopoeic Theory

'The metre was intended to be onomatopoeic', Hardy said. Prior to the twentieth century, there had been relatively little sustained discussion of the function of English metre. Classical definitions—tying a musical or metrical mode to a particular ethos—long permeated English criticism. In 1589, George Puttenham, typically, recommended the alexandrine 'for grave and stately matters' (76). In 1745, Samuel Say concluded: 'We express our Pleasure and our Joy by the *Trochee*, the *Tribrachus* and the *Dactyle*; our Resentment by the *Anapaest* and *Iambic*; while the Slow and Solemn *Spondee* calms the Passions, and composes the Soul' (169). In 1804 William Blake could still follow a theory of decorum: 'the terrific numbers are reserved for the terrific parts' (preface to *Jerusalem*). In most metrical discussion, the functions of metre, to be beautiful, to be expressive, to be appropriate, to be an aid to memory, were taken for granted, while most attention was paid to the nature of metre. (What we see historically is that ideas about the function of metre became more sophisticated as ideas about the nature of metre improved.) A landmark in early discussion is Pope's few lines in *An Essay on Criticism*. After attacking mere emphasis on sound ('Her Voice is all these tuneful folks admire') or syllable-counting ('And ten low words oft creep in one dull line') or conventionality ('With sure returns of still expected rhymes'), Pope then defines the true function of metre which is to (*a*) echo the sense in physical description ('The hoarse, rough verse should like the torrent roar'); (*b*) express feeling ('And bid alternate passions fall and rise'); and (*c*) create music ('The power of Music all our hearts allow'). Pope is only expressing what oft was thought, but the crisp outline and detail of his discussion, and the response it provoked in Johnson, had a great influence on later discussion. Johnson objected to Pope's notion of 'representative metre' and noted that the 'same sequence of syllables' can be used for very different senses ('Life of Pope', also see 'Life of Cowley', *Rambler*, 92, 94, and *Idler*, 60)—much as twentieth-century critics like Richards and Ransom would argue. Johnson thought Pope said that the sound of a line had a meaning independent of its sense; but in fact Pope's formula, 'The sound must seem an Echo to the sense', means that the sense makes us read the sound as confirming the sense, and the sound so identified can be elaborated in delightful ways.[1]

Lord Kames, reviewing the issue in 1762, agreed that there is little 'natural resemblance of sound to signification', but that 'in language, sound and sense being intimately connected, the properties of the one are readily communicated to the other'. This has remained the standard view; and in the 1760s Daniel

---

[1] On Ransom, see Woodring; on Richards, see Wimsatt, 'Concept of Meter', 123 ff. Kames's discussion, cited below, may have influenced Johnson's reconsideration of the issue in his 'Lives' of Pope and Cowley; but Kames is more sanguine than Johnson that word sounds by themselves produce impressions which parallel what is signified. (Johnson does concede in *Rambler*, 94 that some types of verse motion 'may image' the idea of motion expressed.) For modern statements on the nature of verse mimicry, see Stephen Ullmann, *Semantics*, 87; Shapiro and Beum, *A Prosody Handbook*, 15; Hollander, *Vision and Resonance*, 157; Wimsatt, 'In Search of Verbal Mimesis'. More complete treatments are David I. Masson's 'Vowel and Consonant Patterns in Poetry'; Dell Hymes, 'Phonological Aspects of Style: Some English Sonnets'. On the metaphoric functions of metrical rhythm, also see Fussell, *Poetic Meter*, *passim*; Gross, *Sound and Form*, 4, 11, and *passim*; W. F. Knight, 17, 24, and *passim*; and many modern textbooks. The sound of the verses may not have meaningful associations in themselves, though Ransom and Richards's *caveat* has not been fully accepted. But the consensus of the new critics was that verbal sounds can be made to have meaningful associations by the poet who uses sound to 'echo' the sense.

Webb, like Kames, discussed some of the ways in which the poet creates agreement 'between the movement of the verse and the idea which it conveys' (*Remarks*, 33–4). Kames's conclusion in 1762 seems historically correct:

A resemblance between the sound of certain words and their signification, is a beauty that has escaped no critical writer, and yet it is not handled with accuracy by any of them. They have probably been of opinion, that a beauty so obvious to the feeling, requires no explanation. This is an error. . . . (333)

Shelley would eventually suggest a more articulate approach to the subject: 'Sounds as well as thoughts have relations both between each other and towards that which they represent, and a perception of the order of those relations has always been found connected with a perception of the order of the relations of thought' ('A Defense of Poetry', d. 1821, published 1840). What metrical critics needed was a more adequate sense of the abstract structure of metrical form before they could see that form's potential for intersecting with the sense and thus suggesting analogies of sound. They also needed the support of a critical heritage which, from Lessing on, would provide increasingly sophisticated notions about iconic artistic form.

The next major moment in the understanding of metrical function occurs with the Coleridge–Wordsworth discussions. Their definitions are quite general but are informed by a new sense of the metre's role as a structure of meaning in its own right with a somewhat askew relation to the poem's content. The Johnson–Pope dispute concerned the sound symbolism of individual lines; the Wordsworth–Coleridge dispute concerned the functional purposes of metrical structure itself. The conventional view is that for Wordsworth metre is a superadded charm which alternately tempers and raises passion, while for Coleridge metre interacts with poetic meaning in more organic ways. In fact, both men were intrigued by the 'artificial' nature (Coleridge's term) of metrical form, and the way it interacted with meaning and feeling. Wordsworth, in the preface to *Lyrical Ballads*, was intrigued by the 'complex feeling of delight' produced by the 'co-presence' of metre; his enjoyment of 'similitude in dissimilitude' parallels Coleridge's; and Coleridge's discussion of Wordsworth on metre leads to the famous passage on 'the balance or reconciliation of opposite or discordant qualities' (*Biographia Literaria*, XIV). What to do with metre seems a new and interesting problem for the Romantics; but its function could not be settled before its nature was. Once the abstract nature of the metrical frame was defined by the Victorians, it was then possible more adequately to discuss the expressive purposes of that frame. And those purposes could be more consciously developed by poets like Hardy and Hopkins. The metrical form's function as an icon (to use Peirce's and, later, Wimsatt's term) could be seen. But the Victorians themselves did not achieve very much in the way of discussing metrical function. The most extensive discussion was that in George Saintsbury's three-volume *History*, but this was tainted by logorrhoea and generality. More precise discussion had to wait for the new critics.[2]

What we see, as the Victorian age leads into New Criticism, is that discussion of the onomatopoeia of individual words leads to discussion of the onomatopoeia of whole lines and then series of lines. It was one of the triumphs of twentieth-century new criticism to develop this last level of insight. I think that one of the most interesting modern discussions, though it is only the tip of an iceberg of new critical explication of poetic metres, is Arnold Stein's article, 'Structures of Sound in Donne's Verse', published in 1951. There Stein distinguishes various levels of metrical function, some very abstract, some very immediate.[3]

---

[2] The most articulate of late Victorian discussions is the 1895 article by Albert Tolman, 'The Symbolic Value of English Sounds'. Tolman distinguished four dimensions of imitative sound: (*a*) 'muscular imitation', i.e. the muscular effort of speaking a syllable imitates its meaning, as in the word 'horrid'; (*b*) 'muscular analogy', the muscular effort of speaking an entire line parallels the meaning, as in Pope's 'The hoarse, rough verse should like the torrent roar'; (*c*) 'onomatopoeia', as conventionally understood for a word like 'whippoorwill'; and (*d*) 'sound symbolism': in the line about the cicadas which 'sit on the trees and utter delicate sounds', 'the smallness of the vowels symbolizes the smallness of the cicadas' (Tolman, 151). Tolman's phrase, 'sound-symbolism', will become the more favoured way of referring to more extensive kinds of sound analogies.

[3] Stein's four levels, if I can summarize a complex discussion, are: (*a*) a convention of metrical sound promoted for its own sake as an abstract vehicle of expression; (*b*)

Nothing in the Victorian age is as interesting as Stein's essay, though Stein's work, like the new critical work on metrics generally, was made possible by Victorian insight into metrical interplay. None of the Victorian prosodists saw the relevance of the abstract nature of metre to the drama of the poem's meaning, as does Stein. Here the poets were still in advance of the theorists.

When Hardy says 'the metre was intended to be onomatopoeic', his use of the term reflects an interesting philological history in which the meaning broadened to include larger effects. The word had originally denoted the sound analogies of individual words, as in Puttenham where it has nothing to do with metre. Pope illustrated onomatopoeia in the contours of whole lines but without using the word 'onomatopoeia'. Only in the Victorian period does the word itself begin to denote the rhetorical effect of a sequence of sounds: in 1860 Tennyson discussed the onomatopoeia of several lines of Milton (*OED*).

When Hardy uses the word 'onomatopoeic' and applies it to the poem 'On Sturminster Foot-Bridge', he is using it first in Tennyson's sense to describe the sound symbolism of strings of syllables over a sequence of lines. Whether Hardy read the poems of Virgil or Ford Madox Ford, he was on the look-out for good onomatopoeic effects. Ford Madox Ford was surprised at Hardy's enthusiasm for Ford's 'The Brown Owl'. Hardy told him: 'But of course you meant to be onomatopoeic. Ow—Ow—representing the lamenting voices of owls. . . . Like the repeated double O's of the opening of the Second Book of *The Aeneid*' (Ford, 440). In 1868 Hardy tried to take down 'the exact sound of the song of the nightingale' (*Life*, 57). Hardy also read Theodore Watts's description of the *Aeneid*, v. 481, and the way Watts described the contributing metaphor of the metre: 'where the sudden sinking of a stricken ox is rendered by

naturalistic onomatopoeia used for local effects (comprehending elements in Tolman's four levels); (*c*) contributing metaphor, where the plot of the sound is a metaphor for the thought process (i.e. a metrically ambiguous line being analogous to a state of mental uncertainty); and (*d*) an achievement of a music 'that goes beyond the intensity of the statement itself and *completes* it' (34). The last type approaches the abstract dimension of the first type (34), and seems to carry the third type to its final extreme (268) where it becomes 'almost . . . a complete metaphor' (22).

means of rhythm'. Of course this sort of remark was common in classical studies, and examples were common in the *Gradus ad Parnassum*. Hardy's 'The Spring Call' (204) is an example of his interest in local onomatopoeic effects: Wessex blackbirds sing 'pret-ty de-urr!', while Middlesex blackbirds sing 'pehty de-aw!', and Scottish blackbirds sing 'prattie deerh!'.

Even more interestingly, Hardy was referring to the sound symbolism of the verse skeleton as a structure of meaning. In other words he was capitalizing on the Victorian understanding of the abstract dimension of metre and using that understanding for expressive purposes. We shall also see that Hardy explores, more self-consciously than other post-seventeenth-century poets, the mimetic significance of traditional stanza shape—a dimension of mimesis hardly touched on in modern criticism.

Sound Symbolism in Hardy's Stanzas

There is an enormous number of ways in which metre is used in expressive ways in Hardy's poetry. But in order to map out some paths, I have chosen to concentrate on two dimensions of Hardy's metrical expression, perhaps the most interesting dimensions. These can be classified in two ways: one reflecting the defence Hardy made of 'On Sturminster Foot-Bridge', the other reflecting what reviewers called the musical transformation of Hardy's verse. These two types of metrical function are used separately or together in a multitude of interesting ways.

The first type consists in Hardy's use of a stanza form to imitate the various ways in which the mind and the world interact. The varying rhythms of reflection respond to, or are conditioned by, the varying rhythms of motions in the outer world. The mind can be influenced by the steady recurrence of clock chimes, the rhythms of music and dance, the ebb and tide of sea and wind and river, the movement of light and shadow, the motions of riding and walking, and many combinations of these.

To imitate these influences, Hardy develops his theme and associates the metrical form with the theme. He speaks and the rhythms of his speech are pointedly associated with various images. Thus the metrical sound of the poem seems to imitate the way the mind interacts with the experience described.

'On Sturminster Foot-Bridge' (ii. 225) can conveniently illustrate this kind of sound symbolism, because it was an example Hardy cited. The lines, he said, 'were intended to convey by their rhythm the impression of a clucking of ripples into riverside holes when blown upon by an up-stream wind' (*Life*, 390, see also 301, and *Collected Letters*, v. 318). In saying this, Hardy was trying to convince a hostile reviewer of the metre's virtue. Thus in his second *Collected Poems* (1923), Hardy added the word 'Onomatopoeic' as a subscript to the poem's title. The reviewer, Robert Lynd in *The Nation*, 22 December 1917 (in Hardy's collection) wrote: 'One could make as good music as that out of a milk-cart. . . . But even here Mr. Hardy takes us home with him and makes us stand by his side and listen to the clucking stream.' In fact, the reviewer sensed the onomatopoeia, but could not see its point—another example of the cross purposes which characterized Hardy's dialogues with his critics. The poem begins:

> Reticulations creep upon the slack stream's face
> When the wind skims irritably past,
> The current clucks smartly into each hollow place
> That years of flood have scrabbled in the pier's sodden base;
> The floating-lily leaves rot fast.

Elizabeth Hickson gives a conventional scansion of the poem, a combination of hexameter and tetrameter lines, in a duple rising or iambic rhythm ($a^6b^4a^6a^6b^4$ d.r.), presumably:

> Retíc | ulä | tions créep | upón | the sláck | stréam's fáce
> When the wínd | skìms ír | ritä | bly pást

(The blank stress marks represent 'potential accents'.) The sequence of successively stressed syllables suggest the drag of the current, while the sequence of successively unstressed syllables suggest the rapid and skipping wind.

The verse rhythm does not, of course, imitate the current directly. It imitates the 'impression' of the current. 'The song

and water were not medleyed sound . . . Since what she sang
was uttered word by word' (Stevens, 'The Idea of Order at Key
West'). The association suggests that the rhythms of the
speaking voice are influenced by the rhythms of the outer
world, or rather that a common rhythm runs through both:

> It may be that in all her phrases stirred
> The grinding water and the gasping wind;
> But it was she and not the sea we heard.

The specific rhythm associated with the current suggests the
larger rhythm of nature which runs through all things,
including the 'years of flood', the life and death of 'lily leaves',
and the consciousness of the speaker observing and expressing
all this. This last point is the most important. The hostile
reviewer had noted the 'clucking' rhythm but had not realized
its full implications. If the poem were merely an 'imitation of
Nature', there would be no justification for continuing the
rhythm into the final lines:

> And beneath the roof is she who in the dark world shows
> As a lattice-gleam when midnight moans.

The same rhythm persists in the speaking voice as it goes on to
describe a different scene. But the metrical impression, with its
suggestion of natural corrosion and decay, is still there,
functioning as an ironic subliminal counter-current to the last
picture. Nature's pattern of reticulations is replaced by a
human romantic pattern, the 'lattice-gleam' (revised from
'lamp-light' in the manuscript) but the common mortal pattern
can still be felt because of the rhythm.[4] Such interesting effects
are common in Hardy's poetry.

Hickson's scansion is questionable, especially when we get to
the second stanza where the long lines vary from 13 to 14
syllables. Even in the first stanza the hexameter scansion seems
an arbitrary explanation for lines 6, 8, and 9. But the dipodic
organization gives it all a *ratio*.

---

[4] For a study of Hardy's pattern imagery, see Taylor, *Hardy's Poetry*, chapter 2.

Reticulátions crĕep upón the slăck strĕam's făce
   Whĕn the wĭnd skĭms ĭrrităbly păst,
The cŭrrent clúcks smărtly ĭnto ĕach hóllow plăce
That yĕars of flŏod have scrăbbled ĭn the pĭer's sódden băse;
   The flŏating-lĭly lĕaves rŏt făst.

All of the 18 primary accents fall on normally stressed syllables; of the 14 secondary accents, 6 fall on normally unstressed syllables, supporting the dipodic lilt. The dipodic structure, then, applies a very simple principle of binary multiples to the stresses of the line. The minimum metrical structure thus would be a dimeter, the maximum would be as large as the available syllables, following rules for possible pronunciation, could support. The poem's norm usually suggests how many of these potential accents we should actualize, and the reader makes his own choices in performance. We could read the same line as a heptameter ('Reticulátions crĕep upón the slăck strĕam's fáce'), a hexameter, a pentameter, a tetrameter ('Reticulations crĕep upon the slắck stream's fáce'), a trimeter, or even a dimeter ('Reticulátions creep upon the slắck stream's face'). Given the dipodic principle which builds tiers of multiples out of the binary structure of alternating accents, we can see that not only can the metrical norm change (from iambic to anapaestic, say) while the rule (pentameter) is preserved, but the rule for the same line can change as well, from pentameter to tetrameter and so on. And as the rule changes, the norm adjusts itself. Thus the word 'streams' in the above line is a supernumerary accent, part of a spondee, in the hexameter scansion of the line, but becomes a full monosyllabic foot in the heptameter scansion of the line forcing 'face' also to become a monosyllabic foot. A differently defined hierarchy readjusts the spacing on the primary level. Thus the normal speech stress of 'streams' can serve various kinds of multiples.

Therefore we find that the onomatopoeic effect in 'On Sturminster Foot-Bridge' is more complex than we had supposed. 'The impression of a clucking of ripples into riverside holes when blown upon by an up-stream wind', in Hardy's words, is rendered through the shifting multiples of the rhythm, condensing, and expanding syllables like 'streams'. The first two words, 'Reticulations creep', express in a curious way the paradox of metre itself, an abstract pattern or reticulation, which is oddly elastic, which only we human beings see, and which works upon the stream of our language and can constantly generate new forms according to the way the spirit of rhythm moves us:

> Reticulations creep upon the slack stream's face
> When the wind skims irritably past

This sort of sound symbolism, where nature is made to seem to influence the form of the poem, has a complex history: partly in pastoral, beginning with the simple parallel of shepherd's piping with waterfall and windswept pine in Theocritus' first idyll; partly with topographical or local poetry; partly in traditions of the muses or some shaping Spirit of the imagination; partly in late eighteenth-century and Romantic notions of the Aeolian lyre as a model for poetry. The pastoral song that parallels nature in Theocritus is echoed back by nature in Virgil's first eclogue and is influenced by nature in Spenser's 'April' from *The Shepheardes Calender* (where the lay is 'tuned . . . unto the Water's fall'). These three possibilities give richness to Spenser's refrain in 'Prothalamion' (in Palgrave): 'Sweet Thames! run softly, till I end my song'. One strand of the history of pastoral, reaching into Romantic poetry, suggests a tendency for the 'pathetic fallacy' to reverse, so that nature's song moves the song of the singer. The Orphic muse tends to become a sort of Aeolian muse (see below, Chapter 4).

Intertwined with this history of pastoral is the history of English local poetry, traced most exhaustively by Robert Aubin, who interestingly cites at the end of the tradition Handley Moule's 'Praise of Dorset', which Hardy probably read in the 1860s. Hardy said in his *Life* (386), 'It bridges over the years to think that Gray might have seen Wordsworth in his cradle, and Wordsworth might have seen me in mine'— phrasing he took from a 21 December 1916 *TLS* article (Supplement, p. 618) which he pasted in the back of his copy of Gray ('Gray might have seen Wordsworth in his cradle . . .'). Here Hardy was evoking the great tradition of local poetry that influenced him so much: Gray's 'Elegy' and 'Ode on a Distant Prospect of Eton College', Thomson's *Seasons*, Crabbe, Words-worth, *Childe Harold's Pilgrimage*, among others. A great fountainhead of such poetry is Drayton's *Polyolbion*, whose metrical importance we have seen. In what he called his 'chorographical' poem, Drayton strikingly develops a pastoral conceit, foreshadowing Romantic mimesis. An ancient analogy (see Hollander, *Vision*, 280) is here made very concrete. He asks the Muse to inspire him in the following distinctive way: 'as my subject serves, so high or low to strain, / And to the varying earth so suit my varying vein. . . .' Drayton continues at length in lines which Hardy marked in his copy:

As thou hast here a hill, a vale there, there a flood,
A mead here, there a heath, and now and then a wood,
These things so in my Song I naturally may show;
Now, as the mountain high; then, as the valley low;
Here, fruitful as the mead; there as the heath be bare;
Then, as the gloomy wood, I may be rough, though rare.(ii. 13–18)

So led by his topography, Drayton mingles myth, history, and reflection as the landscape determines. The notion that the landscape somehow shapes the movement of the poem is echoed more slightly in Pope's lines and in Denham's 'Cooper's Hill': 'O could I flow like thee and make thy stream / My great example, as it is my theme'. Example and theme were not fully integrated until the Romantic era. An odd document in this history is the poetry of Ossian, with its constant evocation of mythic memory emerging from the mist, rain, and storm, as in one of the many lines which Hardy underlined and copied (Wright, 76–7; *Literary Notebooks*, ii. 465–7, 564–6): 'I behold my departed friends. Their gathering is on Lora, as in the days of other years. Fingal comes like a watery column of mist!' ('The Songs of Selma'). But certainly the greatest influence upon Hardy was that of the 'greater Romantic lyric' (see Abrams, 'Structure and Style', Taylor, 'Hardy and Words-worth'), particularly such poems as 'To Autumn', 'Ode to the West Wind', both of these in Palgrave, and 'Frost at Midnight', where natural processes like the sequences of a season, a storm, or frost, seem to condition the poet's utterance. On the Romantic development of pastoral, John Hollander writes: 'This outdoor music, the blending of the human with the natural, originates in pastoral. . . . But when Romantic poetry begins to take up the pastoral device, something else happens to it conceptually. A dialectic between the inner and the outer, being subject and object, takes it up, and it starts to involve the actualities of the ear and the way in which consciousness itself makes sense of these' (*Vision*, 40–1). Drayton's conceit also came to a rich fruition with the advent of regional fiction. Interestingly, his English topography in the 'Second Song', the song most marked by Hardy, begins with descriptions of the 'Dorsetian fields' (19) and Hardy's central rivers, the Frome 'quitting in her course old Dorcester at last . . .' (82), the Piddle, and the Stour, two descriptions of which Hardy quoted in *Jude*. Drayton's poem had an important influence on Hardy's conception of the 'Wessex Novels', and on the kind of Hardy discussed by John Holloway in *The Victorian Sage* (London, 1953).

Finally the influence of Hardy's neighbour, William Barnes, cannot be overestimated here. Hardy wrote:

Unlike the bucolic poets of old time, Mr. Barnes does not merely use the beauties of nature as a background. . . . Moved by the pervading instinct of the nineteenth century, he gives us whole poems of still life . . . In these the slow green river Stour, with its deep pools whence the trout leaps to the May-fly undisturbed by anglers, is found to be the dearest river of his memories, and the inspirer of some of his happiest effusions. Its multitudinous patches of water-lilies yellow and white, its pollard willows, its heavy-headed bulrushes, are for ever haunting him; and such is the loving fidelity with which the stream is depicted, that one might almost contruct a bird's-eye view of its upper course by joining together the vignettes which are given of this and that point in its length. (*Personal Writings*, 95–6.)

We might say that the tradition of pastoral and local poetry, as modified most extensively by Drayton and then explored in nineteenth-century lyrics, finds a most extraordinary expression in several Hardy poems, where the natural landscape not only haunts the poems but seems to condition their rhythm, imagery, and structure. When Hardy called his poems 'impressions', he had, among other things, this tradition in mind.

The second type of metrical function is more difficult to describe, as the reviewers noted, but accounts for some of Hardy's most distinctive cadences. It is connected with Hardy's interest in onomatopoeic effects, but here the very abstractness of the metre plays a role as metaphor in the poem. Elsewhere I have studied how Hardy's language imitates the way words take shape, express a fleeting reality, and then become archaic. This is consistent with the way, in Hardy, patterns of perception illuminate the world and then grow rigid and skeletal as the world changes. In *Hardy's Poetry, 1860–1928*, I traced how the meditative structure of the major poems follow a development which parallels that of individual words and larger patterns of perception. A speaker meditates within a changing natural setting, and his meditation seems to freeze him in an archaic point of view. A similar plot can be found in Hardy's rhythms, which imitate the way the natural rhythms of language take shape, become fixed, and ossify into a ghostly music of the past. In a sense Hardy is dramatizing the common Victorian idea that prose 'develops' into poetry: prose speech has its vague repetitive patterns which become poetic as they become clearly recurrent, thus metrical (see above, Chapter 1). For Hardy the poem is an archaic crystallization of prose. The poem seems to recapitulate the historical process by which the fresh speech rhythms of the people become the metrical rhythms of the poet, which, in turn, become vulnerable to the resurgence of new speech rhythms. Such vulnerability is not merely the poet's problem. He speaks for the people and shows where their speech is tending. An architectural and obsolescent form is latent in the natural, seemingly spontaneous, rhythms of ordinary speech. Thus where Hopkins sees poetry as revealing the creative vitality latent in speech, Hardy sees poetry as formalizing the obsolescence latent in speech.

'Near Lanivet, 1872' (ii. 168) may serve as a good example of this second type of sound symbolism. It was, Hardy said, an 'often neglected' poem (*Talks*, 24), one of those he liked 'best', 'possibly among the best I have written' (*Collected Letters*, v. 250,

295; vi. 96). The verse form is $a^5b^3a^5b^3$ d.r., a common cross-rhyme stanza though somewhat unusual in its use of the pentameter line. The speaker describes how he and his beloved paused during a walk, and the girl leaned against a post. He cries out: 'Don't.'

> I do not think she heard. Loosing thence she said,
>   As she stepped forth ready to go,
> 'I am rested now.—Something strange came into my head;
>   I wish I had not leant so!'
>
> And wordless we moved onward down from the hill
>   In the west cloud's murked obscure,
> And looking back we could see the handpost still
>   In the solitude of the moor.
>
> 'It struck her too,' I thought, for as if afraid
>   She heavily breathed as we trailed;
> Till she said, 'I did not think how 'twould look in the shade,
>   When I leant there like one nailed.'
>
> I, lightly: 'There's nothing in it. For *you*, anyhow!'
>   —'O I know there is not,' said she . . .
> 'Yet I wonder . . . If no one is bodily crucified now,
>   In spirit one may be!'[5]

[5] The earliest similar metrical form I've found in another poet is that used in Vaughan's 'Corruption'. Interestingly Hardy marked this poem in his 1897 edition of Vaughan and connected it with Wordsworth's 'Intimations' ode (Wright, 15). Vaughan's poem also looks back at a moment of spiritual fall; he speaks of 'man in those early days':

> He drew the curse upon the world and cracked
>   The whole frame with his fall.
>   .   .   .   .   .
> I see Thy curtains are close-drawn; Thy bow
>   Looks dim too in the cloud;
> Sin triumphs still and man is sunk below
>   The center and his shroud.
> All's in deep sleep and night; thick darkness lies
>   And hatcheth o'er Thy people. . . .

In both poems, one about the loss of Eden, the other about the loss of love, there is a journey away from the past which holds the key to the present. In both poems, there is a sense of exile in a scene where Vaughan says 'All was a thorn'. In both poems the light of the setting and the light of joy dim. Hardy had used Vaughan's metrical form earlier in 'A Meeting with Despair' (i. 75), which seems also to contain specific ironic parallels to Vaughan's subject. But by the time he wrote 'Near Lanivet, 1872', the subject of

What interests me here is how Hardy uses 'inexact rhymes and rhythms' to bring 'life into the writing' (*Life*, 105), or, more specifically, how he takes dramatic liberties with the accentual-syllabic expectation to render the experience of a mind caught within what Murry called 'the sombre, ruthless rhythm of life itself' (see below, Afterword). In the stanzas quoted above, the fifth line must be read carefully if it is not to slip into a tetrameter rhythm:

> And wórdless we móved onward dówn from the híll.

The eleventh and thirteenth lines are also difficult:

> Till she said, 'I did not think how 'twould look in the shade,
>
> I, lightly: 'There's nothing in it. For *you* anyhow!'

These lines seem so loose that we remember two common charges made against Hardy: the unredeemed banality of some of his lines and the awkward imposition of the verse pattern on the speech stress. Thus, Thom Gunn (37), like James Southworth earlier (166), has attacked the 'peculiar and uneuphonious' language of the poem. The function of the metre does not perhaps become clear until the last stanza. What a reviewer said of another Hardy poem comes true here: 'there is no line, until you reach the last four, that stops you with its beauty; and you run through the beauty of the last four to reach the end; and then the beauty of the whole takes you and flows back through the whole poem.'

> And we dragged on and on, while we seemed to see
>> In the running of Time's far glass
> Her crucified, as she had wondered if she might be
>> Some day.—Alas, alas!

The penultimate line is at first a metrical enigma, but I think is deliberately placed in its key position. It is seemingly the worst instance of the metre's violent pressure upon the speech rhythms. But Hardy knows what he is doing. The pentameter

cursed human consciousness had become even more complex and desperate. Where Vaughan's poem ends with an image of trumpets and resurrection ('Arise! Thrust in thy sickle'), Hardy's poem confirms a deeper continuing crucifixion.

norm might first lead us to space the line as follows, with the accent marks denoting normal pronunciation:

Her crú | cified, as | she had wón | dered if she | might be.

But we are left with strings of five and three normally unstressed syllables. The pressure is great on us to find some anchors for the metrical accents. So we might impose the following mechanical scheme:

Her crú | cified, as | she had wón | dered if shé | might bé

The necessity to impose this scheme makes us participate in the drama of the poem and unlocks the metrical mystery. Hardy is clearly aware how the experience of readjusting the natural distribution of stresses can contribute to the meaning. Hardy's forcing of the normal speech stress into a rigid pattern suggests a new dimension, a silent partner working through the language. That silent partner, the poem reveals, is the future. The poem is told from the vantage point of 1912 by which time the tragic pattern portended in 1872 has come true. In the re-telling of the original incident, we hear the pattern stealthily imposing its form on the lovers' conversation, and becoming fully revealed in Hardy's late recollection. The misty future of the lovers has become *his* past, which continues to bind him in its belatedly revealed frame. The shift of emphasis from *might* to *be* mimes this process, since what *might* become the case for the lovers *is* the case for the speaker now.

Her crucified, as she had wondered if she *might* be

Her crucified, as she had wondered if *she* might *be*

We might say with one reviewer that the prosody enables the lyric to 'embody at once a brief incident and a long process'.[6]

---

[6] King, 165. In Vaughan's poem the long pentameter line is also occasionally roughened before the resolution of the trimeter line. There is a somewhat similar pentameter crux in Vaughan:

I see | Thy cur | tains are | close-drawn; | Thy bow
Looks dim | too in | the Cloud

Compare also the way the moved accent adds emphasis in Vaughan: 'I see Thy Curtains *are* close-drawn.'

Thus the metrical rhythm which initially served as a conventional frame for the story becomes part of the story's plot and reveals its culmination. When transformed by Hardy's insistent melody, the light and banal phrasing of some of the lines assumes a portentous weight, as the metrical pattern 'emerges'. The decidedly conversational emphasis, 'In spirit one *may* be', is forced into a new mould by our sense of the metrical pressure:

> In spírit one máy be
> In spírit óne may bé

The second reading suggests that the possibility came true: 'In spirit *one* may *be*'. In Hardy, the rhythm by which the speaker had mimed the laboured movement of the lovers, ('we dragged on and on while we seemed to see') now mimes the laboured movement of his own caught mind. (Hardy seems to emphasize this self-paralysed reflection in a revision of the following lines, which originally read:

> Her white-clothed form at this dim-lit cease of day
> 　　Made her look as one crucified
> As I gazed at her from the midst of the dusty way. . . .

In the revision, 'As I gazed at her' became 'In my gaze at her'.)

I think this kind of dramatic abstraction is the most pervasive signature in Hardy's poetry. It operates where a striking use of pattern imagery, word choice, and meditative structure may be missing. It is at the heart of Hardy's 'personal rhythm': 'Once it has been struck out in the open, it is felt as ever present, not alone in his thirty or forty finest poems but almost everywhere in his work.' So I would interpret Theodore Spencer's remark about Hardy's rhythm (as quoted by Blackmur, *Language as Gesture* (New York, 1952), 79). Hardy is sensitive to the way in which the verse skeleton can reveal the pattern latent in any given impression. I would speculate that this fact underlies what Sassoon sensed in Hardy's cadences: 'One hears his everyday voice in so many of his trivial versifyings—& they have an interest quite apart from their poetic de-merits! But what a precious thing to carry in one's head—those cadences of his voice. Posterity may well envy us our incommunicable power of enjoying T.H.'s verse.' (*Best of Friends*, 32.) When Hardy recited, the recitation sounded first perhaps like an ordinary cadence of everyday speech; as he proceeded one sensed the curious interference of the 'versifying'; then 'the beauty of the whole takes you and flows back through the whole poem'. One realizes the progressive disclosure of the controlling form, and the beautiful consistency of the metrical counterpoint with Hardy's ultimate assumptions about poetry and reality. 'Out of the simplest hack phrases of conversation he seems to evolve a magical melody.' The lines seem spontaneously uttered, and then seem flawed and ghostly—haunted by their architecture. The 'diapason' of the metrical spell binds the poet in a past moment of vision while the 'key-creak' of changing reality can still be heard. The poems, Sassoon said, seem 'to acquire a greater intensity now that he is so far away from us', not only because we miss him, as we miss any great poet, but because the poems themselves dramatize how a once living speech becomes a living echo.

In reciting his poems, Hardy may sometimes have followed his own recommendation about reading *The Dynasts*, and imagining it delivered with 'a monotonic delivery of speeches, with dreamy conventional gestures, something in the manner traditionally maintained by the old Christmas mummers, the curiously hypnotizing impressiveness of whose automatic style—that of persons who spoke by no will of their own—may be remembered by all who ever experienced it.' In *The Return of the Native*, Hardy associates this style with a certain kind of authenticity: 'This unweeting manner of performance is the true ring by which, in this refurbishing age, a fossilized survival may be known from a spurious reproduction.' We know that Hardy admired the King James Bible for 'the poetry of the old words':

They translated into the language of their age; then the years began to corrupt that language as spoken, and to add grey lichen to the translation; until the moderns who use the corrupted tongue marvel at the poetry of the old words. When new they were not more than half so poetical. So that Coverdale, Tyndale, and the rest of them are as ghosts what they never were in the flesh.

Again we see the underlying Gothic analogy: 'the years began . . . to add grey lichen to the translation'. For Hardy, the Gothic architect, Gothic style was a brittle fabric of the past, a skeletal reminder of a once lavish vitality.[7]

Several years after Hardy's death, Yeats was to say something which is consistent with the insights of Victorian prosody and with Hardy's feeling for the ghostly dimension of the verse skeleton:

When I speak blank verse and analyse my feelings, I stand at a moment of history when instinct, its traditional songs and dances, its general agreement, is of the past. . . . The contrapuntal structure of the verse, to employ a term adopted by Robert Bridges, combines the past and present. If I repeat the first line of *Paradise Lost* so as to emphasise its five feet I am among the folk singers—'Of mán's first disobédience and the frúit,' but speak it as I should I cross it with another emphasis, that of passionate prose—'Of mán's first disobedience and the frúit'; the folk song is still there, but a ghostly voice, an unvariable possibility, an unconscious norm. What moves me and my hearer is a vivid speech that has no laws except that it must not exorcise the ghostly voice. ('Introduction', 524.)

Hardy's fascination with structural sound symbolism makes his *Collected Poems* an encyclopaedia of metrical expressiveness. The two sorts of sound symbolism we have discussed are closely related in Hardy's mind: their uniting image is the graveyard epitaph evoked in 'During Wind and Rain'. Wind and rain seem to influence the poetic line, which will remain long after as an epitaph of the influence, an impression frozen in verse and in print, and oddly subject to nature's temporal erosion. Similarly, the names at the end of 'During Wind and Rain' are simultaneously outlined and effaced by the rain. Where Shelley said 'make me thy lyre,' Hardy said: 'make me thy epitaph'. To this point, we shall return in Chapter 5.

---

[7] *Dynasts*, preface; *Return of the Native*, II. 4, p. 144; *Life*, 385; on Gothic style in Hardy, see my *Hardy's Poetry*, 48–59; Hardy's notion of intoned verse may have been influenced by his classical background. James Hadley, whose influence on Hopkins is well known, said in 'Greek Rhythm and Metre': 'Even the simplest kinds of verse, the epic hexameter, the dramatic trimeter, were pronounced—they were intended to be pronounced—in a kind of recitative, a sort of semi-musical utterance, with musical accompaniment' (*Essays Philological and Critical* (New York, 1873), 83).

There is another way one might look at Hardy's distinctive relation to the metrical and critical traditions. What Hardy achieved was an accentual-syllabic form peculiarly consistent with a world in flux. For Hardy, the synchronic or patterned time of form and the diachronic or historical time of life are always distinguishable in theory and in actuality. This is the tragedy of form. Forms lose the life which sustains them, life loses the forms which make it memorable. As with life, so too with language. Metrical form and current speech rhythms can never be identical. The patterns language assumes are momentary, they grow out of one time configuration of mind and reality, they grow old, they bind us for a while in their obsolescing frames. In his metrical rhythms, Hardy captures the resonance of the life and death of forms. This resonance constitutes, I think, the 'unique' music of Hardy's poetry. It is a music which is equally consistent with Victorian and modern insights into the basic nature of English accentual-syllabic form.

But perhaps we need not push the implications of Hardy prosody so far. The challenges made by Hardy's critics are often persuasive on a pragmatic level, unless we see the specific intentions behind Hardy's metres. Also Hardy was in a unique position. Coinciding with the insights of the new prosody, his poems were major innovations in using verse form 'iconically' for onomatopoeic purposes, for re-presenting, in Coleridge's fullest sense, the interaction of mind and world in a changing universe. The originality and accomplishment of this aesthetic was not appreciated at the time, and then lost to view as time passed.

Also, Hardy's unique aesthetic was a response to a specific historical situation. When Hardy went to London in the 1860s, he was greeted by a very flourishing poetic tradition but one whose very foundations were being undermined. The 1860s represent a crisis in English poetry.[8] The insights and feelings of the golden 1850s seemed to find adequate fulfilment in that

---

[8] I hope to demonstrate this point, for Hardy at least, in an unfinished manuscript, 'Hardy and the 1860s'. The 1860s is the dark transition decade in Walter Houghton's *The Victorian Frame of Mind 1830–1870* (New Haven, 1957).

decade's amazing renaissance of literary forms; but the darker
implications revealed in the 1860s, as Arnold sensed, seemed to
undermine these forms. The world of Darwin and Marx no
longer rhymed. The novel and the essay of the 1860s were
finding some adequate and permanent forms, but the poem
was in trouble.

Hardy would immerse himself in the accentual-syllabic
tradition, and use forms drawn from ballads, hymns, sonnets,
classical and French forms, and from the entire glossary of lyric
forms in English. He would also invent more rhyming forms
than any other poet in English. His problem was to make those
forms consistent with the insights of the 1860s. He thought of
himself as writing a new poetry, that is, a poetry responsive to
the discoveries that came in 1859 and later. Discussing his first
volume of poetry, he wrote in 1899: 'There is no new poetry;
but the new poet—if he carry the flame on further (and if not
he is no new poet)—comes with a new note. And that new note
it is that troubles the critical waters.' Two years later he quoted
Leslie Stephen: 'We cannot write living poetry on the ancient
model.' Later Hardy copied from a 1906 poetry review the
following sentence: 'We are conscious that real poetry, new
poetry, is in the air of our day, if only somebody could capture
it. Hence we have little patience with echoes of Tenny[son],
hints of Rossetti, and broken lights of Swinb[urne].' In this
same year Hardy wrote to Quiller-Couch: 'I have long
thought . . . that some new, modified, or revived means of
expressing how life strikes us must develope [*sic*] as novels grow
inadequate'. Various of the reviews in Hardy's collection refer
to his poetry as 'a new poetry, new in its music, in its speech', 'a
new music', 'a new kind of poetry', 'a new and rare music',
'strange new music'. Disappointed in an overly 'obvious'
review of his *Selected Poems*, Hardy again wrote to Quiller-
Couch in 1916: 'What a pity that there is no school or science of
criticism—especially in respect of verse.'[9]

[9] *Life*, 300, 308; *Literary Notebooks*, ii. 193; *Collected Letters*, iii. 221; *The Times*, 23 May
1922; *Observer*, 9 Feb. 1919; *Nation*, 13 July 1918; *Bookman*, Jan. 1920; *Observer*, 30 Nov.
1919; *Collected Letters*, v. 194.

One argument of this book is that the newness of Hardy's poetry was its achievement of a kind of mimesis which crowned the great insights of Victorian prosody. We can now trace the growth of Hardy's art of sound symbolism.

4. *The Development of Hardy's Metres*

ONE of the most difficult problems facing a critic of Hardy is how to describe the development of his metrical techniques over a span of decades accounting for over a thousand poems. Such a description would have to take into account a great number of different metrical forms and the great variety of ways in which these metrical forms relate to the poem's subjects. The complexity of the problem, plus the difficulty of dating Hardy's poems, have made the whole subject *terra incognita* for Hardy scholars. Failure to chart such development reinforces the strangely persistent notion that Hardy is a clumsy metrist, fixed in his early liking for mechanical and inexpressive forms.

I have argued elsewhere and I will argue here that our sense of Hardy's development illuminates the individual poems and helps reveal the nature of their achievement. Two of the reviews in Hardy's collection made this point very well: 'Only those, however, who read the whole book can fully realise the extraordinary appropriateness of the diction and the metre to the mood' (*Nation*, 18 December 1909). 'His poems are members of one another; and the reader must know them all before he can properly appreciate any one' (*New Statesman*, 15 December 1917).

We can penetrate this metrical jungle if we follow Hardy's lead. He told his reviewers to consider the identity of his metrical forms, his 'Sapphics', 'heroics', and so forth. He also told them to consider his onomatopoeic purposes. We have seen how these are connected in the sound symbolism of his stanza forms.

Of course, there is an enormous number of other ways in which metre is used by Hardy. And there is also a wide range of quality: 'In the midst of a mass (800 pages good and bad together) of quite ordinary verse and verse experiment, one wants to make a valid selection, implying the history of Hardy's technical biography' (Pound, *Confucius to Cummings*, 326). Within the limits of this book, the kinds of metrical function we have seen used in 'On Sturminster Foot-Bridge' and 'Near Lanivet, 1872' can help us discover a broad range of excellent poems and also trace Hardy's 'technical biography'. He imitates in more complex and interesting ways the impression that the rhythm of the outer world seems to mingle with his thoughts, and the impression that the verse skeleton is the obsolescent form of the experience which it frames. Eventually, in his development, these two types of metrical function co-operate and even fuse. While these developments are taking place, Hardy is also using more complex and interesting metrical forms. When he resumed his poetic career in the 1890s, he proceeded to take traditional verse forms, break them down into their constituents, and recombine them in more compelling ways. If we analyse the technical constituents of Hardy's verse forms, we can see him developing their combinations while exploring their mimetic effects. If we live with Hardy's poems long enough, we can begin to see him as a poet who is remarkably consistent, not in writing the same poem over and over again, but in reusing his material and techniques in new ways. And the new ways are sometimes remarkably inventive and successful.

I do not want to exaggerate the importance of this development. The idea of a developing technique provides a context leading to a fuller appreciation of Hardy's verse. But the analyses done in this chapter can, I hope, stand on their own as individual analyses of skilful metrical poems.[1]

---

[1] An example of a study connecting technical development with increasing excellence is Bate's *Stylistic Development of Keats*. Bate argues that there is 'a general inclination throughout Keats's entire technical development to return gradually, in a degree surpassing most of his contemporaries, to the skeletal integrity of the Augustan line' (86); that is, Keats showed an increasing tendency to avoid the run-on line (81) and feminine endings (120), 'use only traditional variation' (86) and traditional caesura (120), and diminish metrical inversions (112). At the same time, Keats employed 'whatever stylistic means would burden even further the connotative

The poems analysed here have been grouped according to their themes; and within these groupings the poems have been arranged chronologically (in so far as we can determine the chronology).[2] Such grouping reveals that there is a difference in complexity, both thematic and technical, between the earliest and latest examples in each group of poems. Within the groupings, finally, I have attempted to suggest a plot, a progression of metrical discoveries. Given the complexity of poetry, such a plot has to be seen, if seen at all, as going through several byways. We will also see how these groupings develop themes from Hardy's life and novels.

Mere statistics reveal the growing technical complexity of Hardy's experiments. In this Hardy is the reverse of Tennyson, who moved from more to less complicated forms.[3] In groups of poems defined by the same rhyme scheme, we see a consistent development toward greater variety of metrical constituents or line types in poems published later rather than early.[4] Also

intensity and richness of imagery with which his lines are fraught' (91), which also meant including more spondees (120). This combination of restraint and intensity accounts, Bate suggests, for Keats's maturing excellence. But the techniques Bate discusses are more value-neutral than he suggests; some ambiguity lurks under words like 'integrity' of verse form and 'richness' of imagery: such terms slip evaluative notions into descriptive ones. This is not to deny the value of Bate's observations about technique but to acknowledge that they illuminate, but do not entirely explain, poetic excellence. A similar caution applies to this present chapter.

    [2] Despite a wealth of bibliographical detail available from Purdy and others, the dating of Hardy's poems is, surprisingly, an unstudied subject, full of popular and critical misconceptions. Many, if not most, of Hardy's poems can reasonably be assigned to a given year or period. It is true that many Hardy poems derive from old notes and are given many revisions through the years. But when Hardy decided that a poem was substantially written at an early date or was substantially changed by subsequent revisions, he usually indicated this fact in a postscript. A full study of the bibliographical data available and the pattern of Hardy's publications shows that the statistical likelihood of being deceived about the date of a Hardy poem is extremely low.

    [3] See Alicia Ostriker, 'The Three Modes in Tennyson's Prosody', 277, 280; J. F. Pyre, *The Formation of Tennyson's Style*. On the other hand, George Herbert's metrical development seems like a miniature version of Hardy's: see Mary Ellen Rickey, 'Herbert's Technical Development', *JEGP* 62 (1963), 745–60. Rickey notes in Herbert the decreasing production of sonnets, the use of longer more complex stanza forms, more stanza-linking rhymes, and the use of more than one stanza form in a single poem.

    [4] For example, in poems rhyming *ababab*, the 3 earliest poems use only 2 line types (i. 87, 75, 122) and two of these poems are extended hymnal forms; three line types are

Hardy's longer stanzas tend to come later, and within these the more varied rhythms occur later.[5] As in the clock and music groups discussed below, these purely technical groupings reveal the same tendency to take traditional forms, modify them, and transform them into unique forms.

After tracing the technical development and then examining the onomatopoeic series, I discovered that the poems which show the most increase in metrical complexity are those which also illustrate a marked increase in onomatopoeic complexity. 'Haunting Fingers', in the music series, may be the first poem (two other examples being also published about this time) in an *abab* rhyme scheme to use four metrical constituents. 'Logs on

used in a poem (i. 192) published in 1901 and another (i. 294) written in 1902; four line types first occur in a poem (ii. 221) written in 1915. This sort of pattern is true of a large number of other groups with enough examples to be traced (i.e. at least four examples, most with two composed before 1913 and two composed thereafter). Thus: *terza rima* (first, i. 31, last, iii. 239); triplets (i. 269, ii. 382); *aaba* (i. 129, ii. 29); *aabb* (i. 8, iii. 200); *abab* (i. 273, ii. 357); *abcb* (i. 63, iii. 90); *aabba* (i. 80, iii. 272); *abaab* (i. 199, iii. 144); *ababb* (i. 154, ii. 159); *abbaa* (i. 120, ii. 397); *aabccb* (i. 222, ii. 168); *ababab* (i. 75, ii. 221); *ababcb* (i. 183, iii. 156); *ababcc* (i. 82, ii. 236); *abbaab* (ii. 17, iii. 41); *abbacc* (*Dynasts*, 118, iii. 20); *ababccb* (i. 57, ii. 47); *ababbcbc* (i. 14, i. 279). For the complete list of poems, see Hickson's 'Table VI'. Hickson's figures are incorrect for the groups, *aabbcc*, *aaabcccb*, and *abbacddc*, where no pattern is discernible; some minor development, not involving line lengths, can be seen in the groups, *abba* and *abbab*. There is only one apparent counter-example to the pattern of development, namely *ababcdcd*, where four line types occur early (i. 85) rather than late; however if this stanza were considered a combination of two *abab* stanzas, and if these were added to the *abab* group above, the pattern of growing complexity in that group would be confirmed. With this slight exception, we can say that no group of Hardy poems shows a pattern of increasing simplicity like that we find in Tennyson. In tracing increasing complexity, I distinguish line types (trimeters, tetrameters, etc.) from metrical constituents (trimeters and tetrameters being one constituent: see above, 'The Metrical Constituents of Hardy's Verse'). A poem which combines a pentameter with a trimeter is more metrically complex than a poem which combines a tetrameter with a trimeter (like a ballad).

[5] In the *Complete Poetical Works*, 115 poems use stanzas of 9 lines or more (excluding sonnets). Only 8 of these occur in the three volumes published before 1910; 9 were published in the 1914 volume, *Satires of Circumstance*, and the remaining 97 were published in the last four. In these poems using a 9-line stanza, four line types first occur in a poem published in 1917 ('The Figure in the Scene', ii. 216). Hardy's earliest use of the 9-line stanza was the traditional Spenserian stanza in which he translated part of *Ecclesiastes*. In poems using a 10-line stanza, four line types first occur in a poem published in 1925 ('At a Fashionable Dinner', iii. 18). In poems using an 11-line stanza, four line types first occur in poems published in 1925 (i.e. 'Ice on the Highway', iii. 45). In poems using a 12-line stanza, four line types first occur in 'One Who Married Above Him' (iii. 49), published 1925.

the Hearth', in the series imitating light and shadow, is the second *abcb* poem to use three metrical constituents. 'After the Fair', in the clock-time series, is the second *ababab* poem to use three different line types. 'The Five Students', in a series of journey-rhythm poems, is the first *ababcc* poem to use four metrical constituents (five line types). Certain major poems tend to recur in our discussion (here and in Taylor, *Hardy's Poetry*) because in them various evolutionary possibilities are realized at once.

Hardy's most formative decade, artistically, was probably the 1860s and he embodied its sense of history and time and dissolution. He began his poetic career in this decade and also brought to bear his great love of music. Two of his favourite themes for metrical rendering are clock time and music. As Hardy pondered the implications of his most formative decade, he developed increasingly subtler ways to render the corrosion of time, and the spell of music, in his metrical rhythms.

Clock Time

'Her Dilemma', $a^5b^5a^5b^5$ d.r., heroic quatrain.

'A Broken Appointment', $A^2a^5b^5c^5b^5c^5a^5A^2$ d.r., Petrarchan sestet within a refrain.

'An August Midnight', $a^4b^4a^4b^4c^4c^4$ and $a^4a^4b^4b^4c^4c^4$ d.t.r., conventional sestets.

'After the Fair', $a^4b^2a^4b^2a^4b^3$ t.r.

'Copying Architecture in an Old Minster', $a^4b^3a^4c^4b^3$, t.r., with *c* rhyme the same through first four stanzas.

'The Musical Box', $a^2b^4c^4b^4c^4a^2$ d.r., Petrarchan rhyme scheme overall, long metre stanza enclosed within refrain.

In several of Hardy's poems, the sound of a clock seems to influence the metrical rhythm and mingle with the reflections of the speaker. The earliest example is a line in 'Her Dilemma' (i. 16), written in 1866. The poem uses a conventional heroic quatrain like that used in Gray's 'Elegy' or in Arnold's 'Self-Deception' ($a^5b^5a^5b^5$ d.r.). Hardy also marked several heroic quatrains in his *Hymns Ancient and Modern*—some of them about

death (12, 116), or matrimony (578), though with markedly
different implications. The setting of Hardy's poem is, appropriately, a Gothic church in the context of an ancient history
whose roots are irrecoverable:

> The two were silent in a sunless church,
> Whose mildewed walls, uneven paving-stones,
> And wasted carvings passed antique research;
> And nothing broke the clock's dull monotones.

The stanza ends appropriately with an 'imitation', which may
remind us of 'The curfew tolls the knell of parting day'. The
steady iambic pace and the 'spondee' or 'redundant' stress in
the fourth foot imitate the deadness of the clock's regular
stroke. A similar though more striking example of this effect is
from the first stanza of 'A Sign-Seeker' (i. 65) probably written
in the 1890s or earlier and included in *Wessex Poems*:

> I mark the months in liveries dank and dry,
> The noontides many-shaped and hued;
> I see the nightfall shades subtrude,
> And hear the monotonous hours clang negligently by.

What Hardy 'marks' and 'sees' in the first three lines, he 'hears'
in the last; and the metre renders it. But the effect here is hardly
developed as extensively as it is throughout 'A Broken Appointment' (i. 172), probably written after Hardy met Florence
Henniker in 1893 and included in *Poems of the Past and the
Present*.

'A Broken Appointment' takes a somewhat unusual Petrar-
chan sestet and encloses it within a refrain: $A^2a^5b^5c^5b^5c^5a^5A^2$ d.r.
The rhythm of the clock is pointedly established at the outset:

> You did not come,
> And marching Time drew on, and wore me numb.—
> Yet less for loss of your dear presence there
> Than that I thus found lacking in your make
> That high compassion which can overbear
> Reluctance for pure lovingkindness' sake
> Grieved I, when, as the hope-hour stroked its sum,
> You did not come.

In his 1919 article on Hardy, Murry had been struck by the power of the refrain lines: 'one can hear the even pad of destiny' in the poem which 'seems to have been written by the destiny it records' (*Aspects*, 131–2). The effect, Murry said, cannot be 'defined', though the meaning seems 'inevitable'. Rereading the poem for a 1921 lecture, Murry found the onomatopoeic effect clearer: 'we hear the step of marching time, and in the short lines . . . there are the first and the last strokes of the fatal bell' (*Problem of Style*, 25).

There is an interesting relation between the speaker's reasoning and the movement of the clock. The speaker's reasoning leaves the clock behind for several lines, as does the metre with its run-over lines and conversational rhythm. We are plunged into a Petrarchan sestet. Irving Howe, in his interesting analysis, says that these lines 'suffer a certain affectation of grandeur'. I would agree, adding only that Hardy intends a degree of affectation. His feelings hurt, the speaker constructs a highly abstract complaint to the woman who does not love him. He claims he regrets her absence less than a certain moral failure in her. It is as though the speaker were to say: 'It is not altogether an *erotolepsy* that is the matter with me . . . [I]t is partly a . . . craving for loving-kindness in my solitude'. And it is as though Hardy were to comment: whatever her 'virtues . . . it was certain that those items were not at all the cause of his affection for her'. This is, in fact, Hardy's assessment of Jude's sophistry in *Jude the Obscure* (II. 4, p. 115). In 'A Broken Appointment' the clock re-intrudes ironically on the play of sophistry in the dramatic caesura of the penultimate line and in the refrain.

The second stanza launches an even more sophistical argument: only if she loved him could she come, but she should come for the sake of 'human deeds divine'. The hoped-for equation, coming equals loyalty, loyalty equals love, reveals its fallacy in the final lines which refer back to the disabling condition on which the argument was constructed: 'even though it be / You love not Me?'. The play of argument is beautifully integrated with the setting. The last line is a refrain line, like 'You did not come', and refers again to the specific occasion: she is not there, it is late, the clock ticks on. The rhythm of the clock is subliminal in the second stanza, but its steady ironic abrasion can be felt in the abstract phrase, 'a time-torn man' (revised from 'a soul-sad man' in the manuscript). Of course, it is the poem's sense that makes the metrical form seem an echo. The same metrical form can serve other purposes, as this one does later in ' "If you had known" ' (ii. 406). Later Hardy would read in Henry Newbolt's *The Tide of Time*: 'The resemblance of form is often dimmed or entirely concealed by the differences of content—for where the thought-form is new the stanza-form will give, in combination with it, a new variety of its own cadences' (89; see *Collected Letters*, vi. 327).

The theme of time in 'A Broken Appointment' is very well handled, though there is no exceptional complication in the type of sound symbolism employed. In the poems that follow, both the thematic dimension and the technical sound symbolism increase in interest and combine in some very fine examples, notably in Hardy's greatest volume of poetry, *Moments of Vision*, published in 1917. Two poems Hardy wrote at the turn of the century, some time after 'A Broken Appointment', help prepare the way: 'An August Midnight', written in 1899, and 'After the Fair', written in 1902.

'An August Midnight' (i. 184) is written in two conventional sestets. $a^4b^4a^4b^4c^4c^4$ and $a^4a^4b^4b^4c^4c^4$ d.r. They are widely used in the hymns Hardy marked and in many poems he knew, most pertinently perhaps the Palgrave example by G. Sewell (163), 'The Dying Man in His Garden': 'Thou and the worm are brother-kind, / As low, as earthy, and as blind'. Interestingly Hardy would later use the form of the second stanza of 'An August Midnight' when he returned to a similar theme in 'A Necessitarian's Epitaph' (iii. 228). 'An August Midnight' begins:

> A shaded lamp and a waving blind,
> And the beat of a clock from a distant floor:
> On this scene enter—winged, horned, and spined—
> A longlegs, a moth, and a dumbledore;
> While 'mid my page there idly stands
> A sleepy fly, that rubs its hands . . .

The scene is wonderfully Darwinian, with animal and human

personalities jostling together and meeting at this arbitrary point. The second line, in which the 'clock beats trisyllabically' as Harvey Gross says (45), gives an onomatopoeic signature to the prevailing rhythm.

The remainder of the poem is a subtle mixture of iambic and trisyllabic rhythms in which are blended the sounds of the clock and the blind careering of bugs. In so conditioning the speaker's impressions, the rhythm beautifully illustrates the poem's concluding moral.

> They know Earth-secrets that know not I.

He is cut off from 'Earth-secrets', yet they control him in ways he can not withstand.

'After the Fair', $a^4b^2a^4b^2a^4b^3$ t.r., uses trisyllabic rhythms much more strikingly within an extended common hymnal rhyme though with an unusual metrical scheme. The rhythms of the clock are strongly felt, as the dimeter lines broaden out into the final trimeter line of each stanza:

> The singers are gone from the Cornmarket-place
>   With their broadsheets of rhymes,
> The street rings no longer in treble and bass
>   With their skits on the times,
> And the Cross, lately thronged, is a dim naked space
>   That but echoes the stammering chimes.   (i. 294)

All human motion in the poem seems dictated by the waves of the clock: the thronging and emptying of the market place, the dispersal of folk 'into byways and "drongs"' 'as each quarter ding-dongs', the shrill sounds of young voices and the more muted notes of old voices, even the emergence and submergence of different personalities in the third stanza. In the last stanza, the rhythm of the clock seems to stretch back into Roman times and dictate the emergence and extinction of human life through the centuries.

We can see several elements from the above poems combined in examples from *Moments of Vision*. 'Copying Architecture in an Old Minster', $a^4b^3a^4c^4b^3$ t.r., is one of Hardy's most interesting meditative poems. I discussed it extensively in

Hardy's Poetry, 1860–1928. Technically and thematically it climaxes this series of clock poems. The interlinking of these five-line stanzas is the most complex stanza in the series. The stanza seems to bring both hymnal (*abab*) and ballad (*abcb*) rhymes together in a quintet, uses predominantly trisyllabic rhythms, and links the first four stanzas with the *c* rhyme like a recurrent clock chime. As in 'A Broken Appointment', a train of reasoning is punctuated in dramatic ways by the ricochet of that clock chime:

> How smartly the quarters of the hour march by
>   That the jack-o'-clock never forgets;
>   Ding-dong; and before I have traced a cusp's eye,
> Or got the true twist of the ogee over,
>   A double ding-dong ricochetts.   (ii. 171)

As in 'An August Midnight', the persuasive rhythms of the clock merges with other natural movements, like the creeping forth of eve damps and fog. The hypnotic rhythms draw the speaker into a ghostly trance. As in 'After the Fair', the chimes echo through the empty spaces of the Minster and seem to extend back through time and forward into the future:

> Just so did he clang here before I came,
>   And so will he clang when I'm gone
>   Through the Minster's cavernous hollows—the same
> Tale of hours never more to be will he deliver
>   To the speechless midnight and dawn!

The 'Tale of hours' line is an interesting crux line like the one we saw in 'Near Lanivet, 1872'. Coming after the run-over line preceding, the line would seem to read conversationally (with a stress on 'be'):

> Tale of hóurs never more to bé will he delíver

But in pursuit of the tetrameter spacing, we seek to distribute the four stipulated accents in a perhaps mechanical fashion:

> Tale of hóurs | never móre | to be wíll | he delíver.

We may try other schemes to group those tiny supernumerary syllables, but they skip about like the fleeting hours of the poem. There are many other interesting mimetic effects. The poem skilfully shows how the rhythms of thought and the rhythms of the outer world sometimes converge, sometimes diverge, then dramatically reconverge.

'The Musical Box', $a^2b^4c^4b^4c^4a^2$ d.r., is our climactic example from *Moments of Vision*. The musical chimes in the poem act like the clock chimes in the previous poems; their 'thin mechanic air' ticks off the hours and seems to say with a spirit's voice: 'O value what the nonce outpours'. The stanza marks an interesting technical advance over 'A Broken Appointment':

> Lifelong to be
> Seemed the fair colour of the time;
> That there was standing shadowed near
> A spirit who sang to the gentle chime
> Of the self-struck notes, I did not hear,
> I did not see.   (ii. 223)

The stanza looks like a miniaturization of the stanza of 'A Broken Appointment'. It is more complex, however, because here the dimeter lines bracket a traditional hymnal quatrain, while the rhyme scheme as a whole is the same Petrarchan scheme used in the bracketed sestet of 'A Broken Appointment'. Further, there is throughout the poem a continual play between heavily vowelled 'spondees' and a lighter skipping 'anapestic' rhythm (somewhat similar to the effect in 'On Sturminster Foot-Bridge'). Here, such rhythms mime, respectively, the deceptive stability and fleeting hours of the scene. The poem captures beautifully the way the mind, in its 'dull soul-swoon,' resists and yet is caught by the fatal motion of the 'gentle chime':

> At whiles would flit
> Swart bats, whose wings, be-webbed and tanned,
> Whirred like the wheels of ancient clocks:
> She laughed a hailing as she scanned
> Me in the gloom, the tuneful box
> Intoning it.

The poem's stanza has another function which relates it to our second form of metrical function. As in 'Near Lanivet, 1872', the speaker did not realize the pattern until years later. Indeed, he 'did not hear' the chimes at the time and 'did not see' their implications. Thus the stanza form is like the obsolete form of the experience which time only later developed, like a delayed photographic 'development' in the speaker's mind of what once was. More than in the other poems, therefore, Hardy foregrounds the artificiality of the stanza form, an impression created, I think, by the insistent play of the iambic short lines against the iambic-trisyllabic long lines, enclosing them in a kind of boundary. We are more conscious of the architectonic shape of the poem.

Thus the stanza form of 'The Musical Box' has both an immediate dramatic function and a more deferred symbolic function. It imitates how the musical chimes mingle subliminally with the thoughts of the speaker, and also how the passing years have moulded the speaker's consciousness. These two plots reach parallel conclusions: the chimes emerge forcefully at the end announcing that the hours have flown, time emerges also revealing that a lifetime has flown. The speaker is left only with the final fixed rhythm of this life which has taken shape in ways he never anticipated. Time was too subtle. 'The Musical Box' shows us, again, that Hardy's two kinds of sound symbolism are related, in that the blending rhythms of mind and world eventually enmesh the mind ends in a fixed pattern which the world will undermine.

Hardy may have glimpsed this connection in one of his favourite poems, Browning's 'A Toccata of Galuppi's'. Browning is entranced by the melodic rhythms of Galuppi, but the melody is an *old* melody, full of echoes of ancient life, casting its cold glamour over the present: 'In you come with your cold music till I creep through every nerve'. For Browning, the rhythm of Galuppi seems to imitate both the rhythms of contemporary Venetian life and the culmination of those rhythms in a final ancient form. This is the kind of compound mimesis Hardy will develop. In an essay on Barnes, Hardy quoted Browning's description of Galuppi's melody: 'A ghostly cricket, creaking where a house was burned' (*Personal Writings*, 79), and he applied Browning's description to Barnes's poems, which are written 'in a fast-perishing language'. Hardy also copied and underlined Swinburne's words from 'To Victor Hugo' describing those 'Whose hearts . . . *Ache with the pulse of* . . . remembered song'. Indeed we can say that the two types of sound symbolism are always related. Hardy does not imitate the immediate interaction of man and nature, but the remembered interaction. His sound symbolism does not suggest that the poem is organically one with nature, as a Romantic might desire, because the sound symbolism is an echo of what once was, a 'ghostly cricket'.[6]

Such compound metrical imitation brings us to another series of poems in which Hardy imitates the *pulse* of *remembered* song. After considering this second series, I will consider the significance of the technical verse forms upon which Hardy is drawing.

Music

'The Night of the Dance', $a^4b^3b^4a^4b^3$ d.r., expanded hymnal stanza.

'Reminiscences of a Dancing Man', $a^4b^3a^4b^3c^4c^4d^3e^4e^4d^3$ d.r., common eighteenth-century ode rhyme scheme.

'Lines to a Movement in Mozart's E-Flat Symphony', $a^3a^3b^5c^6D^2$ d.r.

'Apostrophe to an Old Psalm Tune,' $a^5b^5b^5c^3c^2$ d.r.

'To My Father's Violin', $a^3a^3b^5c^3d^2c^3d^2d^5b^5$ d.r.

'Haunting Fingers', $a^2b^3a^5b^6$ d.r. and $a^3b^3a^3b^2$ d.r.

Music and metre have a traditional affinity. The development of the musical bar, we remember, contributed over the centuries to a growing understanding of the nature of the metrical structure (see above, pp. 14–15). Also traditional have been metrical imitations of music: Milton's 'Ode on the

---

[6] Hardy referred to Browning's poem again in 1916, in letters to Gosse and Quiller-Couch: 'I am beginning to feel chilly & grown old, as the man did who listened to the Toccata of Galuppi's. . . .' (*Collected Letters*, v. 183, 149.)

Morning of Christ's Nativity', Dryden's 'Song for St. Cecilia's Day', Cowley's 'A Supplication', Dryden's 'Alexander's Feast', Collins's 'The Passions: An Ode for Music', and Shelley throughout—all appear in Palgrave and create intricate metrical forms to express varying musical moods. Such poems reflect the kind of development John Hollander and others have traced: from music's one-time association with heavenly power and the harmonies of the spheres, to its post-Renaissance century secularization and association primarily with the stirring of passion, to its Romantic association with an enchantment of imagination, and finally the spell of memory: 'Music, when soft voices die, / Vibrates in the memory', as in Palgrave's last selection. The 1860s saw Browning's publication of one of several influential poems in this tradition and perhaps his most metrically interesting one: 'Abt Vogler'. The poem, whose dipody we have cited, is indeed about music and about poetry, seen as building parallel creative structures. Browning reminds us that theories of music parallel theories of imagination. The demythologizing and secular reinvestment of music finds a parallel in the history of imagination—in which history Hardy holds an important place.[7]

Despite Hardy's well-known musical interests, his contribution to the tradition of musical sound symbolism in verse has never been evaluated. Yet in musical form Hardy found a good equivalent for his notion of the imagination, a fundamental framing force of the human mind, locking it in obsolescing dreams. It is Hardy, not Browning, who represents the most sustained and inventive creator of metrical versions of musical effects. His ambition was to create a new music which would be consistent with the assumptions he began to share with others in the 1860s. His metres would create a melody imitating the flawed enchantment of the human mind in a world of 'concordia discors' as Hardy described it in 'Genitrix Laesa' (736) and other cosmological poems. He was deeply impressed with the paradox with which Browning struggled (in DeVane's paraphrase): 'If such a fugue or such a suite would catch a soul

---

[7] See Taylor, *Hardy's Poetry*, 159 n. 3.

heavenward once, why does it fail now? Isn't perfection always perfect?'[8] The Aeolian harp will vibrate differently in Hardy's 'The Darkling Thrush': 'The tangled bine-stems scored the sky / Like strings of broken lyres (i. 187).

The *Life* says that as a child of four Hardy 'was of ecstatic temperament, extraordinarily sensitive to music, and among the endless jigs, hornpipes, reels, waltzes, and country-dances that his father played of an evening in his early married years, and to which the boy danced a *pas seul* in the middle of the room, there were three or four that always moved the child to tears'. 'This sensitiveness to melody . . . remained with him through life'. Not only dance melodies, but ballads, hymns, operas, and especially music hall songs filled his mind and haunted his years: 'The history of the theatre is to my mind nothing to the history of the concert room', he said after referring to Arthur Sullivan and J. Stainer. He felt that the most marvellous song in English music was 'Should He Upbraid'. The last lines of this song, 'And dance and play / And wrinkled care beguile' are repeated over and over in different combinations in an entrancing crescendo—'drawing out the soul of listeners', Hardy said of one performance, 'in a gradual thread of excruciating attenuation like silk from a cocoon'. This kind of rapturous susceptibility characterizes many of the characters in Hardy's novels. In his later years, Hardy would attempt to hunt down melodies he remembered, just as he tried to relocate Gothic architectural patterns he loved. As late as 1918 when he was 78, he danced and played the fiddle; he continued to play the fiddle through his eighties. In 1920, he visited Exeter Cathedral and quoted 'Abt Vogler': 'Felt I should prefer to be a cathedral organist to anything in the world. "Bidding my organ obey, calling its keys to their work, claiming each slave of the soul".' In 1923 he would recall 'words and music he had heard played at his father's house, Upper Bockhampton, when he was a boy'. In this year Hardy told Brennecke: 'I can't even always fathom quite the charm of the ancient church musicians about here. They serenaded me

---

[8] William C. DeVane, *Browning's Parleying* (New Haven, 1927), 274.

with some old tunes the other evening. That sort of thing
carries me back to the fifties—even to the forties.'[9]

What Hardy had long experienced, and often described in
his novels,[10] he would dramatize with increasing skill in the
rhythms of his poems. For his first volume of poetry, he mused:
'Title:—"Songs of Five-and-Twenty Years". Arrangement of
the songs: Lyric Ecstasy inspired by music to have precedence'
(*Life*, 243). As his poetic models he prized poets like Swinburne, Shelley, and Poe, admired for their lyric ecstasy and
musicality. We noted that earlier several of his early reviewers
acknowledged a transforming sense of melody in his poems—a
claim which has grown increasingly more mysterious over the
years as we have forgotten what the old tunes were like. A rare
modern testimony is that of Philip Larkin who said he was
immediately struck by the 'tunefulness' of Hardy's verse.[11]

[9] *Life*, 15, 16; *Collected Letters*, ii. 283; *Life*, 118, 123, 219; Mardon, 9; *Life*, 404;
Brennecke, *The Life of Thomas Hardy* (New York, 1925), 5; Mardon, 14. The *Life*, 269,
claims that Hardy's last dancing, 'at any rate on the greensward', took place in 1895.
Mardon says that often after 1918 she played duets with Hardy: 'he on his fiddle, and I
on the piano, his mind was clearly back in the days of childhood' (12). Brennecke has a
good section on music in Hardy's novels, 146–69. Other important references to
Hardy's musical interests and background are in the *Life*, 18, 20, 22 ff, 43, 376, 404.
Hardy's music books have been microfilmed in *Original Manuscripts*, Reel 10. Other
musical influences are cited by Elna Sherman in her two articles of 1940. Also, see
Pinion, *Hardy Companion*, 187–93; Ruth Firor, ch. 9; John Udal, chs. 11 and 12; Carl
Weber, *Hardy Music at Colby* (Waterville, Me., 1945).

[10] Characters subject to the spell of music in Hardy's novels include Cytherea in
Desperate Remedies, VIII. 4, p. 155; Elizabeth and Henchard respectively in *The Mayor of
Casterbridge*, VIII, p. 52, XXXVIII, p. 312; Tess in *Tess of the d'Urbervilles*, XIII, p. 107;
Car'line in 'The Fiddler of the Reels'. On music in Hardy's novels, see also Evelyn
Hardy, *The Countryman's Ear*, ch. 8: 'What began as a healing, reconciling influence, to
draw parted lovers together, has ended as a destructive force, driving them apart' (77).

[11] ('Philip Larkin Praises the Poetry of Thomas Hardy'). Hardy spoke of Shelley as
'our most marvellous lyrist', the 'poet he loved', 'the highest-soaring among all our
lyrists', 'that greatest of our lyrists' (*Life*, 17, 131; *Personal Writings*, 81; *Collected Letters*,
vi. 101). Also see *Personal Writings*, 107; Pinion 213–14, and Phyllis Bartlett's articles
(Gerber and Davis, No. 2457–8). On Hardy and Poe, see above. For Swinburne's
influence on Hardy, see the Metrical Appendix below; also Pinion, 486, 208; Purdy,
106; *Life*, 287, 349, and *passim*; Brennecke, 5; *Collected Letters*, iv. 25 ('What music there
was in S!'). H. A. T. Johnson points out that Palgrave's preface emphasizes those very
poetic qualities which Hardy is supposedly in rebellion against: 'Thomas Hardy and
the Respectable Muse'. Hardy is often described as an anti-musical poet in the
tradition of Donne—a description which ignores the cunning metrical lyricism of both
men and their love of intricate stanza patterns. Though Hardy owned and marked a

I cannot hope to trace the wealth of metrical echoes of music in Hardy's poems. In the following group of poems, I am interested in what seems to me the major examples of his musical mimesis, which he conveys with increasingly complex metrical means. Hardy admired Amy Lowell's 'After Hearing a Waltz by Bartok': 'the metre & rhythm keep up the beat of the waltz admirably', Hardy said (*Collected Letters*, v. 67). We shall see exploration of such effects in his poems.

'The Night of the Dance', $a^4b^3b^4a^4b^3$ d.r., may be the earliest of the series (it is placed in a grouping of mostly 1860s poems). It is written in an expanded hymnal stanza (with trisyllabic substitutions). Its theme imitates that of Tennyson's 'Move Eastward, Happy Earth'. The last stanza creates a lovely lilting imitation of 'the sound / Of measures trod to tunes renowned'. This stanza reveals what everyone has been waiting for:

> That spigots are pulled and viols strung;
>   That soon will arise the sound
> Of measures trod to tunes renowned;
> That She will return in Love's low tongue
>   My vows as we wheel around.   (i. 282)

The rhythm pauses nicely on occasion over a redundant stress as a sort of syncopation: 'That She will return in lóve's lòw tóngue'.

'Reminiscences of a Dancing Man', $a^4b^3a^4b^3c^4c^4d^3e^4e^4d^3$ d.r., is a memory of the dance halls Hardy used to attend in the 1860s.

> Who now remembers Almack's balls—
>   Willis's sometime named—
> In those two smooth-floored upper halls
>   For faded ones so famed?

copy of Donne's poems, given him in 1908 (Evelyn Hardy, 303), Hardy makes no reference to Donne before the 1920s when Donne was receiving much critical attention. In 1922 Hardy compares Donne's 'view of the world, the grave, etc.' to his own (Wright, 15; also see Bailey, 211). About the same time, in the *Life*, Hardy compares Donne's indifference to publication to his own indifference (*Life*, 49). When Clement Shorter suggested that Crabbe's 'realistic style' (apparently Hardy's term for it) was the most potent influence on Hardy's writings, Hardy denied that it had any influence on his poetry (*Collected Letters*, v. 294).

> Where as we trod to trilling sound
> The fancied phantoms stood around,
>   Or joined us in the maze,
> Of the powdered Dears from Georgian years,
> Whose dust lay in sightless sealed-up biers,
>   The fairest of former days.   (i. 266)

The next two stanzas mention 'gay Cremorne,' 'Jullien's grand quadrilles,' and, in archaic language, 'those crowded rooms / Of old yclept "The Argyle" '. Early in the *Life*, Hardy refers to 'the Argyle Rooms and Cremorne' and a 'fascinating quadrille' whistled by one of his fellow architects: it 'remained with Hardy all his life, but he never could identify it' (*Life*, 34). Seven chapters later Hardy remembered another incident of hearing the quadrille played by an organ-grinder and now identified it as 'possibly one of Jullien's'.[12] Between the time these incidents occurred and the time he dictated the *Life*, Hardy wrote 'Reminiscences of a Dancing Man' (about 1895).

The tetrameter-trimeter pattern of 'The Night of the Dance' is here expanded into something much more elaborate. The form had been used before in Thomas Campbell's 'The Last Man' and, with the exception of one line, Gray's 'Ode on a Distant Prospect of Eton College' (Palgrave, 158). The rhyme scheme is common in eighteenth-century odes. Hardy associates the stanza form with the turns and counter-turns of the remembered melody. The poem dramatizes not only the dance rhythms, but also dramatizes how those rhythms have become an old echo. Though the dancers are gone, the melody continues in the verse and thus seems to hover in the void like a spectral theme. We can hear the dance stumble a little in the redundant stress, 'Whose dust lay'.

The poem is another example of Hardy showing in a poem what he asserts in a novel. In *The Hand of Ethelberta*, the violinist Christopher finds himself caught up in a crescendo of dancing: 'Watching the couples whirl and turn, advance and recede as

---

[12] *Life*, 123. In 1887 Hardy was 'obsessed by an old French tune of his father's, "The Bridge of Lodi" ', and tried unsuccessfully to locate it (*Life*, 195–6). In 1889 he also heard by chance an old man sing 'How oft, Louisa': 'though Hardy searched for him afterwards he never saw him any more' (*Life*, 219).

gently as spirits . . . and lullabied by the faint regular beat of their footsteps to the tune', he falls into a kind of spell and the noises of the dance come 'to his ears like voices from those old times when he had mingled in similar scenes. . . .' (iv, pp. 40–1). The puns in Hardy's poem also support the metrical suggestion of what is at once the brief incident of the dance and the long process of the years: 'fancied' as romantic lovers at the time, later called up in fancy by Hardy; 'the maze' of the dance pattern, the maze of the years; 'powdered' with rouge, powdered with grave dust. In Campbell's poem, the world disintegrates but the human spirit is immortal; Gray's theme is more consistent with Hardy's: 'Who foremost now delight to cleave / With pliant arm, thy glassy wave?'

Interesting as 'Reminiscences of a Dancing Man' is, Hardy's technical effects become much more elaborate in his subsequent efforts, while the thematic dimension grows as well. 'Lines to a Movement in Mozart's E-Flat Symphony', $a^3a^3b^5c^6D^2$ d.r. (begun in 1898 and finished some time before its publication in *Moments of Vision*), is an elaborate metrical imitation of Mozart's third 'Menuetto' movement. The stanza is an unusual one, using lines of four different metrical types; and the stanzas are linked by the *b* rhyme and the refrain. The form imitates the sense of expansion and contraction suggested by Mozart's rhythm, the rapid strokings of the violins imitated in the penultimate long line, their slowed climaxes caught in the final short line:

> Show me again the day
> When from the sandy bay
> We looked together upon the pestered sea!—
> Yea, to such surging, swaying, sighing, swelling, shrinking,
> Love lures life on.   (ii. 195)

The poem attempts to recapture, in order, first the 'time', then the 'day', then the 'hour', and finally the 'moment' of the early love experience. This plot seems to suggest that less and less is recapturable, though the intensity of the remembering increases. Indeed, the expansion and contraction of the rhythm imitates the way memory expands in the mind while the reality of the past contracts into nothing. While the poem is obviously an experimental *tour de force*, the consistency of metre and theme certainly keeps it from being the monstrosity of 'heavy engineering' it has been called.[13]

'Apostrophe to an Old Psalm Tune,' $a^5b^5b^5c^3c^2$ d.r., written in 1916, fully exploits this irony of how a musical memory grows in power while one's personal world contracts with age. The poem is about an old psalm tune that Hardy hears again and again over the years: 'But you waylaid me. I rose and went as a ghost goes. . . .' The stanza form, an unusual one, mimes the way in which the strong hymn rhythms invade the speaker's reflections long after he thought he would never hear them again:

> Much riper in years I met you—in a temple
> Where summer sunset streamed upon our shapes,
> And you spread over me like a gauze that drapes,
> And flapped from floor to rafters,
> Sweet as angels' laughters.   (ii, 163)

The reflective pentameter lines give way to the stronger rhythms of the two final short lines. The feminine ending of the fourth line makes us come down hard on the first syllable of the fifth line, thus giving the fifth line a falling rhythm, creating a powerful counterpoint (in Hopkins's sense). The indentation and parallel later lines make the last line a dimeter with an interesting choriambic swing or dipodic potential: 'Sweet as angel's laughters'. Over the years, with changes in church ritual, the tune is stripped of its elaborations and then, the speaker thinks, abandoned. (For this development, Hardy in line 12 blames, among others, William Henry Monk, musical editor of *Hymns Ancient and Modern*.) The middle stanzas mime this loss of music by means of run-over lines and jarring

---

[13] Davie, 17. The poem may also be influenced by the contraction-expansion effect in Shakespeare's sonnet, 'That time of year thou mayst in me behold'. Shakespeare describes his age first as a time of year, then a twilight of a day, then a glowing of a fire. This plot of diminishment and intensity may have had an influence elsewhere on Hardy's poetry: see examples below, and also Taylor, *Hardy's Poetry, passim*. On this poem, Hardy wrote to a friend: 'To make the verses fit the music you would of course have to repeat the words of the last line but one in each verse' (*Collected Letters*, v. 237).

caesuras in the pentameter lines. But the rhythmic pattern seems forcefully to reassert its control in the last two stanzas when the speaker hears the psalm again:

> So, your quired oracles beat till they make me tremble
> As I discern your mien in the old attire,
> Here in these turmoiled years of belligerent fire
> Living still on—and onward, maybe,
> Till Doom's great day be!

Two years later, a younger colleague of Hardy's will describe in a famous poem 'the great black piano appassionato'. The effect is beautifully rendered by Hardy in the dimeter boom on 'Doom's' and 'dáy be'. We will see this again in Hardy: a falling rhythm urged by the metre and superimposed on a normally rising rhythm. In other words, we hear the expected metrical rhythm intersect more strongly with the reflective rhythms of the speaking voice—somewhat like the power of music casting its spell on the consciousness of the speaker. That more hypnotic rhythm was, we discover, implicit all along; and its echo now seems, eerily, to outlive life itself.

Hardy's '1867' notebook shows that he was intrigued by a story called 'Thrond' in Björnstjerne Björnson's *The Bridal March* (London, 1884, trans. Rasmus Anderson), which Hardy owned. The story describes how a young fiddler is enchanted by the melodies he plays; and several details of the story parallel details Hardy used in descriptions of his characters and even of himself in the *Life*. One of the sentences Hardy copied from the story reads: 'He played till his father told him he was fading away before his eyes' (*Literary Notebooks*, ii. 462). By the time Hardy wrote 'Apostrophe to an Old Psalm Tune' in 1916, he was able to create the metrical equivalent of losing one's life in a musical reverie.

'To My Father's Violin', $a^3a^3b^5c^3d^2c^3d^2d^5b^5$ d.r., was also written in 1916 and represents an important advance in this series in the sense that the stanza form is more intricate than the earlier examples. When a reviewer condemned the poem for its pessimism, Hardy wrote: 'Such is English criticism, and I repeat, why did I ever write a line!' (*Life*, 410). More recently the poem has been attacked as an example of those requiring 'a wrenching of the word to fit the obviously artificial patterns'. The nine-line stanza is a beautifully intricate and original one, using three different metrical line lengths. Some of the poem's images are like those in 'The Night of the Dance' and 'Reminiscences of a Dancing Man'. As we read the poem, we grow to realize that we are re-enacting, metrically, a nearly visual equivalent of the complex rhythms which Hardy's father played on the violin:

> And, too, what merry tunes
> He would bow at nights or noons
> That chanced to find him bent to lute a measure,
> When he made you speak his heart
> As in dream,
> Without book or music-chart,
> On some theme
> Elusive as a jack-o'-lanthorn's gleam,
> And the psalm of duty shelved for trill of pleasure   (ii. 186)

Though Hardy asserts its extinction, this elusive music continues to bind his mind. The violin is a 'tangled wreck' and the violinist inhabits Virgilian 'glades / Of silentness'. Yet the father's rhythms live on in the son's metrical impressions.

'To My Father's Violin' also does something else which is quite remarkable. We saw that in 'Apostrophe to an Old Psalm Tune' the last line suggests a falling rhythm which seems to counter the prevailing rising rhythm. In 'To My Father's Violin', Hardy often seems to suggest both rhythms at once, as though one rhythm were ghosting the other:

> Ìn the gállery wést the náve
> Bùt a féw yards fróm his gráve,
> Dìd you, túcked benéath his chín, tò his bówing
> Guìde the hómely hármonỳ
> Óf the quíre

Indeed, the more we read the poem, the more we find the rising scansion, $a^3a^3b^5c^3d^2c^3d^2d^5b^5$ d.r. (Hickson's reading), change to a falling scansion, $a^4a^4b^6c^4d^2c^4d^2d^6b^6$ d.f. We have seen how the dipodic possibility supports such transformation. This delightful syncopation makes for some funny emphases on the initial minor syllables: 'Élúsive ás a jáck-o'-lánthorn's gléam', 'Wéll, yóu, can nót, alás' / Thé bárrier óverpáss'. (To be consistent, Hickson should have made her dimeter lines monometers; they become dimeters in the falling scansion.)

'Haunting Fingers' (ii. 357), subtitled 'A Phantasy in a Museum of Musical Instruments', $a^2b^3a^5b^6$ d.r. and $a^3b^3a^3b^2$ d.r., also mimes the spectral effect of music outliving death. Here those spectral melodies have become entities which interplay with and answer each other. Numerous instruments 'speak' in the poem. Such is Hardy's version of the sequence of musical passions explored in the grand *Golden Treasury* odes by Dryden and Collins; in Hardy the music has been reduced to old echoes lingering around museum instruments: 'And they felt past handlers clutch them.' For the first time in this series, two different stanza forms, one for the instruments, one for the narrator, are used in a sort of musical dialogue. The poem thus uses more metrical elements than we have yet seen, four metrical line lengths and two different stanzas, one very unusual, the other perhaps borrowed from an anthology (see Metrical Appendix) and also used by Hardy in a poem of a somewhat similar theme, 'The Selfsame Song' (ii. 367). Hardy praised 'Haunting Fingers' as one of the two best poems privately printed by Florence Hardy (the other poem being 'Voices From Things Growing in a Churchyard', Bailey, 449). In 'Haunting Fingers', the abandoned instruments themselves speak (like the pots in Fitzgerald's *Rubáiyát*):

> 'My keys' white shine,
>   Now sallow, met a hand
> Even whiter. . . . Tones of hers fell forth with mine
> In sowings of sound so sweet no lover could withstand!'

> And its clavier was filmed with fingers
>   Like tapering flames—wan, cold—
> Or the nebulous light that lingers
>   In charnel mould.

The stanzas draw on previous techniques and play against each other. We seem to hear a sort of dialogue between the metres.

The growing echo of the first stanza—swelling from a two-beat to a six-beat line—is answered by the ballad-like stanza, with its starker narration dwindling to a dimetre. The first stanza builds to a long-line climax, like the effect of the penultimate lines in 'Lines to a Movement in Mozart's E-Flat Symphony'. The second stanza resolves the complex pattern of the first into a more traditional and recognizable pattern, the kind of resolution we saw carried out in each stanza of 'Apostrophe to an Old Psalm Tune'.

The evolutionary development of this series properly ends with 'Haunting Fingers'. But to see how much Hardy has learned since 'The Night of the Dance', we might note the very late 'Concerning Agnes', written about 1927. 'Concerning Agnes', $a^5b^2a^5b^2c^5c^6$ d.r., is also about a remembered dance but in a new and more complex stanza, which uses a conventional sestain rhyme-scheme but adapts to it three different line lengths, including a successful hexameter. The personal speech rhythm is skilfully mingled with the turns of the dance whose beat is felt as a ghostly background:

> I could not, though I should wish, have over again
> > That old romance,
> And sit apart in the shade as we sat then
> > After the dance
> The while I held her hand, and, to the booms
> Of contrabassos, feet still pulsed from the distant rooms.   (iii. 215)

The dance rhythm has become a static repetition in the mind of Hardy who is himself, like Agnes, 'one of the Nine grown stiff from thought'. The reversed beats and the redundant stresses drag beautifully against the dance rhythm:

> Yes. She lies white, straight, features marble-keen,
> Unapproachable, mute, in a nook I have never seen.

In writing this series of time and music poems, Hardy achieves increasingly subtle effects. Thematically we see an increasing complexity in the rhythmic impressions which the verse tries to simulate. The relation of this rhythm to the movement of consciousness becomes more interesting. And there is a growing self-consciousness that the metrical form symbolizes an experience which has assumed a fixed shape over a long period of years.

Technically, we see that metrical hints in the early poems may be taken up and become the shaping principle of later poems. There is more variation in the iambic pattern, more play with trisyllabic rhythms, and more exploitation of the relation between falling and rising rhythms. And there is more experiment with the varying lengths and resonances of vowels and consonants. We also see a tendency to begin with conventional verse forms in the earlier examples, then experiment with unusual rhythms within conventional or slightly modified conventional forms, then invent new forms altogether. The development of the stanza form is accompanied by the use of a greater variety of line lengths. In later work, Hardy will use more and more pointed contrasts of very short and very long lines. These technical advances enable Hardy to achieve more diverse onomatopoeic effects. These principles remain fairly constant in the other series of onomatopoeic poems we will examine.

Thus in the clock series, Hardy begins with the heroic quatrain in 'Her Dilemma', plays with a Petrarchan stanza enclosed within a dimeter refrain in 'A Broken Appointment', turns to some conventional sestets in 'An August Midnight', then a more unusual sestet with more trisyllabic rhythms in 'After the Fair'. The penultimate poem in the series gives us an ingenious fusion of ballad and hymn stanzas in a quintet with the rhyme interlocked throughout the first four stanzas ('Copying Architecture in an Old Minster'); and the final poem, 'The Musical Box', fuses a Petrarchan rhyme scheme with a hymnal stanza and a dimeter refrain.

Again, in the music series, Hardy begins with an extended hymn stanza in 'The Night of the Dance', goes to a longer but still conventional stanza form in 'Reminiscences of a Dancing Man', then tries a *tour de force* in the unusual stanza and rhythms imitating Mozart in 'Lines to a Movement in Mozart's E-Flat Symphony'. The last three examples in the series are increasingly more complex stanzas used for matured imitative purposes: the five-line stanza with three line types in 'Apostrophe to an Old Psalm Tune' with a suggestion of competing rhythms, the nine-line stanza with three line types in 'To My Father's Violin' with a full elaboration of competing rhythms, and two different stanzas using a total of four lines types in 'Haunting Fingers'.

Growing complexity of effect and technique can also be seen in the following series of 'wind and water' poems. These have a special relationship to a favourite theme of Hardy's imagination. They suggest a new dimension of influence over the shaping of the poem. In tracing this series, and the next, we shall also keep track of the growing technical complexity of the stanzas.

Wind and Water

'Friends Beyond', $a^8b^4a^8$ d.f., *terza rima*

'The Homecoming', $a^7a^7\ b^7b^7b^7b^7$ d.r.

'The Souls of the Slain', $a^3b^2c^2c^3a^4b^2$ t.r.

'In Front of the Landscape', $a^5b^2c^5d^2e^5b^2$ t.f.

' "I found her out there" ', $a^2b^2b^2a^2c^2d^2c^2d^2$ t.d.r.

'The Phantom Horsewoman', $a^4b^2c^2b^2c^2b^2c^2a^2a^4$ d.t.r.

'During Wind and Rain', $a^3b^3c^3b^3c^3d^3a^4$ d.r.

Hardy's novels often refer to the mingling of human voices with sounds of water and weather. *A Pair of Blue Eyes* compares the 'purl and babble' of tongues on a busy London street to the 'ripple of a brook'. In *Far from the Madding Crowd*, the 'low peal of laughter' from Troy's friends is 'hardly distinguishable from the gurgle of the tiny whirlpools outside'. In *The Hand of Ethelberta*, songs mingle with the wind and tide 'till not the echo of a tone remained'. Winterborne's delirious monologue in *The Woodlanders* is 'like that we sometimes hear from inanimate nature in deep secret places where water flows, or where ivy leaves flap against stones'. In *The Return of the Native* he uses the image with a new insight: 'her articulation was but as another phrase of the same discourse as theirs'. This insight underlies the many examples of such imagery in *The Woodlanders*. Later *The Well-Beloved* asserts the metaphysical implication: voices and ocean mingle so that 'the articulate heave of water and the articulate heave of life seemed but differing utterances of the selfsame troubled terrestrial Being. . . .' Thus, Hardy echoes the Aeolian tradition, particularly as pondered by Coleridge in 'The Eolian Harp': 'And what if all of animated nature / Be but organic Harps diversly fram'd. . . .'[14]

An important image in the Romantic development of pastoral and local poetry is that of the 'Aeolian harp', an image developed in Thomson, Collins, Ossian, and explored most powerfully in the 'correspondent breeze' of the Romantic poets (see Abrams's 'Correspondent Breeze', Grigson, Erhardt-Siebold, and Hollander's 'Wordsworth'). The 'timely utterance' in the 'Intimations' ode may be an Aeolian visitation, as the context suggests, and evoke an obsolete meaning of 'timely', i.e. 'keeping time or measure', as in the Spenser example cited by the *OED*: 'many Bardes . . . Can tune their timely voices cunningly' (*Faerie Queene*, v. 3). Hardy also knew (in Palgrave) Gray's 'The Progress of Poesy' with its opening address, 'Awake, Aeolian lyre', which, Gray rebuked a reviewer, was *not* to be confused with the Aeolian harp. Instead, Gray said, he was evoking Pindaric phrases, 'Aeolian song, 'the breath of Aeolian flutes'. Oddly enough, this mistake may have been a vehicle whereby the Aeolian harp became conflated with the Pindaric muse, and so became a prime analogy for the Romantic imagination. In any event, the analogy was a very powerful one for Hardy. C. H. Salter's list (pp. 103–4) of many examples in the novels where the wind plays music in the trees and reeds should be combined with the others just cited, where human voices blend with the sound of wind and water. Hardy's 'The Darkling Thrush' alludes to the broken strings of the lyre, and 'On the Way' refers punningly to the wind as 'lyre' and

---

[14] *Pair of Blue Eyes*, XIII, p. 148; *Far from the Madding Crowd*, XI, p. 100; *Hand of Ethelberta*, XLIII, p. 392; *Woodlanders*, XLII, p. 377; *Return of the Native*, I. 6, p. 61; *Well-Beloved*, II. 13, p. 138; *Tess*, XVIII, p. 154; *Tess*, XX, p. 165. Also, when Tess and Angel are courting, they 'were never out of the sound of some purling weir, whose buzz accompanied their own murmuring' (XXXI, p. 247).

liar. We shall see that 'During Wind and Rain' is a natural development of 'Ode to the West Wind'. When Hardy defined his poems as based on 'impressions', he may have had in mind a passage, parts of which he copied from Arnold's essay, 'Maurice De Guérin'. The first and last underlinings are by Hardy:

To make *magically real and near* the life of Nature, and man's life only so far as it is a part of that Nature, was his faculty; a faculty of naturalistic, not of moral interpretation. This faculty always has for its basis a peculiar temperament, an extraordinary delicacy of organisation and susceptibility to impressions; in exercising it the poet is in a great degree passive (Wordsworth thus speaks of a *wise passiveness*); he aspires to be a sort of *human Aeolian harp*, catching and rendering every rustle of Nature. (*Literary Notebooks*, i. 129.)

'Welcoming every impression without attaching itself to any,' Hardy added, summarizing the passage. He may also have been influenced by Arnold's more sinister version of the Aeolian lyre in *Empedocles on Etna*, ll. 78–86:

> Hither and thither spins
> The wind-borne, mirroring soul,
> A thousand glimpses wins,
> And never sees a whole. . . .

One of Hardy's most typical lyrics is one in which the speaker meditates within a changing physical setting; his meditation is subtly formed by that setting and yet grows blind to its slow changes until they reveal themselves in various surprising ways. The interruption of a meditation comes to stand for other kinds of interruption—the interruption of a dream or an habitual point of view or a way of living or life itself. The smaller interruption 'recapitulates' the larger. I have already traced the development of this central lyric in *Hardy's Poetry, 1860–1928*. Here I wish briefly to consider the skilful metrical dimension of such poems, especially where the metre dramatizes the changing relation of mind and setting. Again Wallace Stevens serves to remind us of the precise nature of such sound symbolism:

> It may be that in all her phrases stirred
> The grinding water and the gasping wind;
> But it was she and not the sea we heard.

Though it is Hardy's speakers we hear, nevertheless they are caught up within a larger world—of settings, of words, of rhythms. But unlike the Stevens speaker, perhaps, the Hardy speaker is not quite 'the single artificer of the world / In which she sang'. He and his poem seem to be orchestrated by a pervasive force of Nature—thus enacting in themselves what the novels assert.

'Friends Beyond', $a^8b^4a^8$ and continuing as *terza rima*, is the most interesting example from *Wessex Poems*. It was one of Swinburne's favourite poems from this volume.[15] The speaker listens to the voices of ghosts who lie in Mellstock churchyard. Hardy incorporates an unusual combination of octometer and tetrameter lines in the traditional form. Within the form we may hear a variation on an extended ballad measure $(a^4b^4c^4d^4b^4)$, here stretched into longer lines and given an hypnotic trochaic rhythm:

> They've a way of whispering to me—fellow-wight who yet abide—
>   In the muted, measured note
> Of a ripple under archways, or a lone cave's stillicide   (i. 78)

The rhythm, in its typically falling fashion, seems to catch up the normally unstressed syllables in its tide, forcing us to redistribute accents on the initial syllables of the line. For example:

> Thé|ve a | wáy of | whíspering | tó me— | féllow- | wíght who | yét
> a | bíde—

on the analogy of the first line,

> Wílliam | Déwy, | Tránter | Réuben, | Fármer | Lédlow | láte
>   at | plóugh

There is also a strong dipodic motion:

> They've a wäy of whispering tö me—fellow-wïght who yét abïde—
>   Ïn the müted, méasured nöte

15 Swinburne does indeed list 'Friends Beyond' as one of his favourites—i.e. 'the frontispiece poem'—contrary to an assertion in Taylor, *Hardy's Poetry*, 161 n. 10.

Overall, the metrical form captures the elusive wave-like motions of ghostly whispering which blends with 'the muted, measured note / Of a ripple under archways'. Such sound symbolism becomes a subliminal force in the rest of the poem until it intrudes ever so subtly into consciousness in the last line: 'And the Squire, and Lady Susan, murmur mildly to me now'.

'The Homecoming', $a^7a^7\ b^7b^7b^7b^7$ d.r. (i. 303), written in 1901, is a more boisterous experiment in such sound symbolism. The form draws on the ancient fourteeners, here arranged in couplets and quatrains. The rhythm seems to imitate the windy bluster of the storm. This bluster carries over into the rhythm of the husband's gruff speeches to his reluctant bride. Occasional lines end in a striking use of monosyllabic feet which make the human scene pointedly echo the natural scene:

> *The wind of winter mooed and mouthed their chimney like a horn,*
> *And round the house and past the house 'twas leafless and lorn*
>
> 'But my dear and tender poppet, then, how came ye to agree
> In Ivel church this morning? Sure, thereright you married me!'
> —'Hoo-hoo!—I don't know—I forgot how strange and far
> 'twould be,
> An' I wish I was at home again with dear daddee!'
>
> *Gruffly growled the wind on Toller downland broad and bare,*
> *And lone | some was | the house | and dark; | and few | came | there.*[16]

Or, in a dipodic scansion:

> *And lonesome was the house and dark; and few came there*

Wimsatt and Beardsley would call this a substitution of an iamb ˘´ for a dipodic foot ˘´˘`, i.e. 'came there' instead of, say, 'are coming there'. Notice that in our reading of the line, that which is a redundant stress, part of a spondee in the heptameter reading of the line ('and few / came there'), is the dipodic secondary accent in the dipodic reading. The metre shimmers with different structural possibilities, and supports the mimesis.

---

[16] *Complete Poetical Works*, i. 203, my scansion marks; Hardy's example may also recall certain children's chants: 'Sally's in the kitchen doing a bit of knitting / In came the bogey man and out popped she' (quoted by Ian Robinson, *Chaucers Prosody* (Cambridge, 1971), 49).

The tempo of the storm and the rhythm increases as the argument grows in intensity, until both begin to subside. It was some time before writing 'The Homecoming' that Hardy read Theodore Watts's remarks on metrics (see above, Chapter 2) and his description of Virgil's *Georgics* (ii. 441) where 'the gusts of wind about a tree are rendered as completely as though the voice were that of the wind itself'.

'The Souls of the Slain', $a^3b^2c^2c^3a^4b^2$ t.r., written in 1899, is a major landmark in the series of Hardy's meditative poems. It is the first to elaborate a complex physical setting at the beginning of the poem and then emphasize a dramatic return to the setting at the end. Entranced by the rhythm of the tides, the speaker is drawn into a vision of the ghosts of dead soldiers. The second stanza describes the tides in a form which seems to imitate their ebb and flow. The same form will also be used to describe the speeches of the ghosts and the broodings of the speaker.

> No wind fanned the flats of the ocean,
>   Or promontory sides,
>   Or the ooze by the strand,
>   Or the bent-bearded slope of the land,
> Whose base took its rest amid everlong motion
>   Of criss-crossing tides.   (i. 124)

Auden particularly applauded 'The Souls of the Slain' for its ability to teach the young poet 'how to make words fit into a complicated structure and also, if he is sensitive to such things, much about the influence of form upon content' ('Literary Transference'). The sestet is an original one using two constituents: tetrameter-trimeter lines and dimeter lines. Hardy may have noticed the same rhyme scheme in Swinburne's *Atalanta in Calydon* ('She has filled with sighing the city'). The rhythm is much more subtly varied than in the first two poems in this series, with its pointed use of iambic and anapaestic interplay and its greater relishing of assonance. It captures nicely the interweaving motions of mind and ocean. The poem also suggests their ominous separation, though some of the implications are unclear. 'The Homecoming' was written two years later but is comparatively a comic *tour de force*. Later we shall see Hardy combine the long dramatic line of 'Friends Beyond' and 'The Homecoming' with the more complex stanza form of 'The Souls of the Slain'.

Nothing in *The Dynasts* is as interesting as 'The Souls of the Slain' (but also see below, Note on *The Dynasts*). After *The Dynasts* Hardy returns to the structural possibilities of lyric mimesis. 'Lyrics and Reveries' in *Satires of Circumstance* seems to be a transitional grouping connecting the period of *The Dynasts* with the period of 'Poems of 1912–13'. The grouping contains several good examples of the wind and water series. The most striking of these is 'In Front of the Landscape', placed at the beginning.

'In Front of the Landscape', $a^5b^2c^5d^2e^5b^2$ t.f., makes the oceanic rhythm a clear metaphor for a psychological state of mind. It imitates the speaker '[p]lunging and labouring on in a tide of visions . . . Through whose eddies there glimmered the customed landscape. . . .' These rhythms seem to swell up out of the past and fix the speaker in a vivid yet belated vision:

> Thus do they now show hourly before the intenser
> 　　Stare of the mind
> As they were ghosts avenging their slights by my bypast
> 　　Body-borne eyes,
> Show, too, with fuller translation than rested upon them
> 　　As living kind.　　　　　　　　　　　　　　　　(ii. 7)

The stanza form is an unusual one, using two strongly contrasting constituents. The *d* rhyme of one stanza becomes the *b* rhyme of the next—which is a more complicated form of the interconnecting system we have seen before. The trochaic and dactylic rhythms of the long lines and the choriambic roll of the short lines catch the wave-like motion of the reverie, receding, advancing, until the interlocking rhymes climax in the last two: 'looms' and 'tombs' (which changes the rhyme scheme of the last stanza to *abcded*). Metaphors of the relation between ocean and reverie are spread throughout the poem. The poem suggests a rich paradox. The great formative moments of our past catch us up in their tides, but the

implications of those moments do not become clear until years
later.[17]

There is a sense of rough experimentalism about 'In Front of
the Landscape', with its surreal and crowded images, its
confusing multiplicity of tenses, and its exaggerated rhythms.
In the following series we see Hardy express the paradox of
belated realization in more delicate and economical ways. The
effect he achieves depends partly on his use of the dimeter line.
In December 1912, he achieves an extremely beautiful effect by
using only dimeter lines.

'"I found her out there"', $a^2b^2b^2a^2c^2d^2c^2d^2$ t.d.r., uses an
eight-line stanza form very close to the one Byron uses
($a^2b^2a^2b^2c^2d^2c^2d^2$, d.r.) in 'When We Two Parted'. In the poem
Hardy confesses that he had taken Emma away from the
beauty of Cornwall. Yet though the poem asserts the loss of
Cornwall sea and storm, their rhythms persist and reverberate
ironically through the poem. They dramatize the paradox
defined above, for they stand both for the shaping form of the
past which casts its spell over the present and for the reality of
the present which breaks through the archaic spell:

> Yet her shade, maybe,
> Will creep underground
> Till it catch the sound
> Of that western sea
> As it swells and sobs
> Where she once domiciled,
> And joy in its throbs
> With the heart of a child.   (ii. 51)

The anapaestic final line seems to release the clipped pent-up
rhythms of the preceding iambic-anapaestic lines.

'"I found her out there"', the rhythm implies, is as much
about Hardy as about Emma. The implication is made fully
clear in one of Hardy's great metrical syntheses, 'The Phantom
Horsewoman', written in 1913.

---

[17] For a recent extensive discussion of this poem, see J. Hillis Miller, 'Topography
and Tropography in Thomas Hardy's *In Front of the Landscape*', in *Identity of the Literary
Text*, ed. Mario Valdes and Owen Miller (Toronto, 1985).

'The Phantom Horsewoman', $a^4b^2c^2b^2c^2b^2c^2a^2a^4$ d.t.r., combines two constituents in the longest stanza form yet used in this series. The onomatopoeic effect is a subtle one, the roll of sea and reverie much more muted than in 'In Front of the Landscape'. In the first stanza the dimeter lines describe the lover while the framing tetrameter lines reflect the detachment of the narrator. Eventually we grow to realize that the sea-haunted lover *is* the narrator who is haunted by his past. The play of seeming detachment against deep-rooted involvement is suggested by the rhythm with its play of reflective iambic rhythms against hypnotic anapaestic rhythms. The internal dimeter lines which begin as a sestet, *ababab*, are then linked back with the enclosing tetrameter lines by the final rhyme. The distinction between narrator and lover deftly breaks down:

> Queer are the ways of a man I know:
> He comes and stands
> In a careworn craze,
> And looks at the sands
> And the seaward haze,
> With moveless hands
> And face and gaze,
> Then turns to go . . .
> And what does he see when he gazes so?　　(ii. 65)

With its muted rhythms and formal columnar structure, the poem comes across as the last phantom figuring of a powerful emotion. With superb economy and grace, the poem fuses Hardy's two types of metrical function, naturalistic imitation and mechanic echo. The romantic rhythms of mind and ocean have staled into this present and belated form of realization.

'During Wind and Rain', $a^3b^3c^3b^2c^3d^3a^4$ d.r., is both metrically and structurally the climactic poem of Hardy's most mature collection, *Moments of Vision*. The poem, Hardy said, was 'possibly among the best I have written' (*Collected Letters*, vi. 96). Several techniques used in earlier poems in the series are synthesized here. Like the early poems, it builds a unique stanza form out of a traditional ballad form. It uses the same line types as 'The Souls of the Slain'. The final tetrameter line of each stanza, with its striking spondaic rhythm reinforced by assonance, describes the advance of the storm which grows in intensity.

> They sing their dearest songs—
> He, she, all of them—yea,
> Treble and tenor and bass,
>   And one to play;
> With the candles mooning each face. . . .
>   Ah, no; the years O!
> How the sick leaves reel down in throngs!   (ii. 239)

Just as the storm interrupts the speaker's reflections, the tetrameters jar the rising, almost nursery-rhyme rhythms of the earlier lines. As in 'The Phantom Horsewoman', a line in the tetrameter frame rhymes with a line in the body of the sestet and thus suggests a hidden fate. The metrical pattern beautifully imitates the way the speaker's mind leaves reality, is drawn sharply back to it, in a series of cycles such as we saw in 'The Souls of the Slain'. Here, however, the implications are beautifully clear.[18]

This series of metrical imitations properly climaxes with 'During Wind and Rain'. To see how far Hardy has come since 'Friends Beyond', we can cite 'Voices from Things Growing in a Churchyard', written in 1921. Hardy regarded this poem and 'Haunting Fingers' (the climactic example in the music series) as the two best poems privately printed by Florence Hardy.

'Voices from Things Growing in a Churchyard', $a^4B^2a^4a^4c^4c^4D^2D^2$ d.r., is a fine and subtle experiment. It exploits the refrain more than the earlier poems and is also the first to play the falling rhythms of the refrains against the rising rhythms of the stanzas. The voice of each ghost is given a distinctive rhythm and a distinctive series of vowel and

---

[18] The first stanza may echo a children's rhyme popular in Dorset: 'One to stop, / Two to stay, / Three to make ready, / And four away'—quoted by Udal, 384. Late in life, Hardy's 'conversations on music were . . . exclusively and extensively on Dorset folk music, popular ballads and the like'—according to Vera Mardon, 21. In Taylor, *Hardy's Poetry*, ch. 1, 'The Development of Hardy's Meditative Lyric', is also about the development of 'During Wind and Rain', in so far as it draws on the meditative frames and imagery of earlier poems.

consonant sounds. These reflect both the ghost's personality and the natural object into which each ghost, like Drummer Hodge, has grown. The flightiness of Fanny Hurd, grown into daisy shapes, is imitated in the run-over lines and the skipping /t/ sounds which smooth out into /s/ and /ei/ sounds:

> These flowers are I, poor Fanny Hurd,
>   Sir or Madam,
> A little girl here sepultured.
> Once I flit-fluttered like a bird
> Above the grass, as now I wave
> In daisy shapes above my grave,
>   All day cheerily,
>   All night eerily!   (ii. 395)

For repressed Bachelor Bowring, who finally became 'a dancer in green as leaves on a wall', the heavy /n/ sounds and the steady iambic rhythm are teased by trisyllabic substitutions. So with the other ghosts. (A similar characterization is made of each instrument in 'Haunting Fingers'.) Punctuating the ghost speeches is the contrasting rhythm of the refrain lines. The first refrain expresses a common impersonality used in correspondence; the second expresses a more sinister impersonality. In Hardy's world, to communicate with ghosts is to become ghostly. The poem's last stanza pictures the speaker in a graveyard,[19] caught up in an unearthly spiritual-material hum:

> —And so these maskers breathe to each
>   Sir or Madam
> Who lingers there, and their lively speech
> Affords an interpreter much to teach,
> As their murmurous accents seem to come
> Thence hitheraround in a radiant hum,
>   All day cheerily
>   All night eerily!

In using the same stanza form to express his own reflections,

---

[19] Again, as in 'During Wind and Rain', the concrete setting is an important element—which Hardy emphasized by adding 'in a Churchyard' to the title when the poem was collected. Later, as Pinion points out, he confused the title with that of Gray's elegy (*Life*, 413).

Hardy '[a]ffords an interpreter much to teach'. Hardy is on one side of the grave, the ghosts on the other, and the difference, we grow to realize, is not so great. Both are caught, in their respective ways, in the changing churchyard scene while their minds—or spirits—are still fixed in their forms of 'interpreting'.

There is another important series of Hardy poems in which the number of metrical constituents increases and is used in more subtle ways. This series is perhaps Hardy's most intriguing achievement in metrical mimesis and is one of the great achievements of traditional accentual-syllabic verse. Hardy imitates, through his stanza patterns, the way in which the mind's motions interact with the motions of light—firelight, moonlight, sunlight.

Light and Shadow

'A Cathedral Façade at Midnight', $a^5b^3a^5b^3c^5c^5c^5$ d.r.

'A Commonplace Day', $a^3b^7a^3b^7b^3$ d.r.

'The Shadow on the Stone', $a^3a^4b^4a^4c^4c^4d^4c^4$ d.t.r.

'Old Furniture', $a^4b^4a^4b^4b^2$ d.r.

'Logs on the Hearth', $a^4b^2c^5b^4$ d.r.

'The Pedigree', $a^3b^5c^5a^7d^3c^7b^6$ etc., d.r. (first stanza)

This kind of imitation also climaxes a gradual development of the image in the novels. We have seen how Hardy in the novels compares human voices to sounds in outer nature—which then leads in the poems to his metrical imitation of 'wind and water' effects. In *Tess of the d'Urbervilles*, Hardy writes the most interesting of these sorts of passages—which leads to an even more interesting sort of metrical imitation. Angel is just about to notice Tess:

One day, however, when he had been conning one of his music-scores, and by force of imagination was hearing the tune in his head, he lapsed into listlessness, and the music-sheet rolled to the hearth. He looked at the fire of logs, with its one flame pirouetting on the top in a dying dance after the breakfast-cooking and boiling, and it seemed to jig to his inward tune; also at the two chimney crooks dangling down from the cotterel or cross-bar, plumed with soot which quivered to the same melody; also at the half-empty kettle whining an accompaniment. The conversation at the table mixed in with his phantasmal orchestra. . . . (xviii, p. 154)

Angel does not know it but his course is about to converge with Tess's, 'as surely as two streams in one vale'. The coalescence of multiple effects in the above passage is striking. It brings together an imagined music, overheard voices, and the flickering of the fire (and the quivering of the soot, which is perhaps what Coleridge meant in 'Frost at Midnight' by the 'film . . . on the grate' whose flutter becomes a companionable form to meditation). When Hardy turns to poetry, he will gradually perfect a metrical form which will embody this companionable form.

'A Cathedral Façade at Midnight', $a^5b^3a^5b^3c^5c^5c^5$ d.r., was written in 1897 and is the earliest example in this series. The rhyme scheme is fairly common, an extended heroic sestet, but Hardy substitutes two trimeter lines. The metrical form is associated with the way the light begins '[i]nch by inch thinly peeping', skimming the statues' 'scantly toe, breast, arm', until in the last three lines of the second sestet the light fully extends and 'the stiff images stood irradiate'. The tentative effect of the trimeter lines with their feminine rhymes changes to the strong confirmation of the concluding pentameter lines with their three climactic rhymes:

> The lunar look skimmed scantly toe, breast, arm,
> > Then edged on slowly, slightly,
> To shoulder, hand, face; till each austere form
> > Was blanched its whole length brightly
> Of prophet, king, queen, cardinal in state,
> That dead men's tools had striven to simulate;
> And the stiff images stood irradiate.　　　　　(iii. 9)

As in 'On Sturminster Foot-Bridge', a play of skipping rhythms against strongly stressed rhythms contributes to the effect, here the light peeping and then growing stronger. The metrical icon ingeniously fits the theme of the death of religion: the light blanches the old statues and exposes the obsolescence of the coded creeds. The 'verse skeleton' mimes this photograph-like development of an exposed and ancient image, blanched by the light of the present.

'A Commonplace Day', $a^3b^7a^3b^7b^3$ d.r., achieves a more complex effect and is another major landmark in Hardy's maturing integration of mind and setting:

> I part the fire-gnawed logs,
> Rake forth the embers, spoil the busy flames, and lay the ends
> Upon the shining dogs;
> Further and further from the nooks the twilight's stride extends,
> And beamless black impends.   (i. 148)

The quintet stanza uses a common rhyme scheme, but Hardy incorporates unusual heptameter lines. The poem describes how the flickering of firelight, and the long shadows thrown by the setting sun, blend with the speaker's reflections. As the sunlight dies, hope dies, and as the fire flares up, regret flares up. The rhythm, with its unusual alternation of trimeters and heptameters, achieves a supple imitation of this inner and outer movement. The aural effect seems dictated by the lighting, the lines expanding and contracting, slowing and skipping, to match the play of light which leads the speaker's thoughts. The poem also suggests that the poet's awareness awakens belatedly, it is like a 'dying flame' nourished by darkness, it flares up just before extinction, it leaves us with a 'verse skeleton' glowing like a pattern of dying embers:

> And undervoicings of this loss to man's futurity
> May wake regret in me.[20]

Hardy's greatest examples of the mimesis of light were, as we might expect, written after 1912.

'The Shadow on the Stone' might be mentioned in passing. It was written between 1913 and 1916. This unique stanza,

[20] Hardy's striking use of 'undervoicings' may have been influenced by Mrs Humphrey Ward's *Robert Elsmere* (London, 1888), a book which Hardy referred to and quoted from (*Collected Letters*, i. 176; *Literary Notebooks*, i. 211). Rose's nature 'had been touched at last . . . by the piercing under-voices of things—the moral message from the world' (VI. 39, p. 490).

$a^3a^4b^4a^4c^4c^4d^4c^4$ d.t.r., an octave combining two *aaba* quatrains, uses one constituent, the tetrameter-trimeter line, but with great suppleness. The stanza imitates a subtle meditative movement that harmonizes with the flickering lights and shadows. These last, Hardy says, 'shaped in my imagining / To the shade that a well-known head and shoulders / Threw there when she was gardening' (ii. 280). The poem ends with this 'apparition' of Emma carefully maintained by Hardy ('Nay, I'll not unvision / A shape which, somehow, there may be'), a strategy carried out in the imagery and the delicately preserved verse skeleton of the poem.

'Old Furniture', $a^4b^4a^4b^4b^2$ d.r., is a wonderful example from *Moments of Vision*. The stanza was used by Swinburne in 'Félise' and in *Hymns Ancient and Modern*, 629, marked by Hardy and also listed by him in the back of his hymnal. The form is also used by Hardy in 'At Castle Boterel', and what C. Day Lewis said of that poem may apply to this: 'The variety of emotion is equalled by that of stanza-form: it is as though the diverse moulds had been preparing through a lifetime, and now those scenes from the past ran freely into them, each recognizing its own' (171). Indeed the metrical mould comes out of a lifetime of poetry which began a half-century earlier with the sapphic stanza, of which this stanza form is a descendant (see Metrical Appendix, p. 258 ff.). The poem shows the interplay of tradition and experiment so characteristic of Hardy. He uses many trisyllabic substitutions and arranges the stanza's rhythm and shape on the page to echo the influence of firelight on the mind. The images cast by the fire are described 'Glowing forth in fits from the dark, / And fading again, as the linten cinder / Kindles to red at the flinty spark, / Or goes out stark'. The firelight dallies over the 'relics of householdry' and is associated with other spectral motions: the dallying of ghostly hands on the furniture, the tentative touches of a moth, like fingers touching a clock, the dancing of fingers on an old viol, the glowing forth of faces (as in the first stanza of 'During Wind and Rain'). The relic of vitality in all these motions is caught beautifully in Hardy's classic image of a mirror series: 'I see the hands of the generations . . .

> Hands behind hands, growing paler and paler,
>   As in a mirror, a candle-flame
> Shows images of itself, each frailer
> As it recedes, though the eye may frame
>   Its shape the same.   (ii. 227)

The narrator is himself like a dying man ('He should not continue in this stay, / But sink away'), the very rhythms of whose speech are the echoes of older rhythms, the fading images of images—though the ear may frame their shapes the same. The verse frame is like the final skeleton of his own consciousness.

Another of Hardy's most accomplished experiments in structural light symbolism is 'Logs on the Hearth' (ii. 232), written in 1915, probably about the same time as 'Old Furniture'. The poem was written in memory of Hardy's sister, Mary, who had died in the autumn of 1915. 'She was almost my only companion in childhood,' Hardy said (*Collected Letters*, v. 135).

'Logs on the Hearth' builds on a ballad rhyme scheme ($a^4b^2c^5b^4$ d.r.), which changes to a hymnal rhyme scheme in the last stanza: $a^4b^2a^5b^4$ d.r. This is an extraordinary quatrain, using three metrical constituents, the pentameter, the tetrameter, and the dimeter line:

> The fire advances along the log
>   Of the tree we felled,
> Which bloomed and bore striped apples by the peck
> Till its last hour of bearing knelled.   (ii. 232)

It imitates with greater economy and subtlety than 'A Commonplace Day' the flickering of the fire, its extension and recession. The speaker remembers when the log was a tree limb: it had bloomed and bore apples, it had been climbed by him and his sister, it had been pruned where the bark is charred—as it is charring now in the flame. In other words, the tree's life had interacted with the lives of Hardy and his sister, and its life and decline has paralleled theirs. The poem thus climaxes a theme which intrigued Hardy in *The Woodlanders*, where old John South has an uncanny obsession with the tree outside his window. When it is cut down, his 'whole system seemed paralyzed by amazement' and he died (xiv, p. 122).

In the poem Hardy, as is often the case, makes his own the experience he narrates about one of his characters in the novel. The tree log's combustion now interacts with and promotes Hardy's memory, which is all that remains of their life together. The fire 'advances along the log' and seemingly along these stages of the speaker's life, which combust like burning photographs. Just before they turn to ash, however, they are seen most vividly in the flame:

> My fellow-climber rises dim
>   From her chilly grave—
> Just as she was, her foot near mine on the bending limb,
>   Laughing, her young brown hand awave.

The fourth line ('Làughing, | her yóung | bròwn hánd | awáve') wonderfully fuses brown hand and wood log—the redundant stress in 'young brown hand', emphasizing the effect. The flickering trisyllabic rhythm of the third line yields to the emphatic strongly stressed rhythms of the final line. The emotionality is beautifully integrated with the imagistic pattern. The image of his sister, with him on the bending tree limb, rises like a film from the flame. The momentary patterns of the verse led by the flame and projecting its image seem to hover in the void. Surely this is one of the great poems of English literature.

'The Pedigree', $a^3b^5c^5a^7d^3c^7b^6$ d.r. in the first stanza, was written in 1916 and is Hardy's most technically complex achievement in the metrical imitation of light and shadow. It is among the most prosodically varied of Hardy's poems, written in four septets and a final octave, each stanza with a different rhyme scheme and metrical form.

> I bent in the deep of night
>   Over a pedigree the chronicler gave
> As mine; and as I bent there, half-unrobed,
> The uncurtained panes of my window-square let in the watery light
>   Of the moon in its old age:
> And green-rheumed clouds were hurrying past where mute and
>   cold it globed
>   Like a drifting dolphin's eye seen through a lapping wave.

>   (ii. 197)

The poem uses three constituents, trimeters, pentameters, and the longer lines, hexameters and heptameters. The varying rhythm models the subtle patterning of moonlight (an effect attempted also but not so successfully in 'Honeymoon-time at an Inn' (ii. 262)). Behind the clouds the light 'globed / Like a drifting dolphin's eye seen through a lapping wave', and the light extends into Hardy's room. It illuminates the hieroglyphs of Hardy's pedigree whose tangles twist into a face tokening toward the window. The window becomes a mirror reflecting images behind images as in 'Old Furniture'. The theme here is more fully developed, for those images are Hardy's own ancestors patterning and forestalling 'every heave and coil and move I made / Within my brain, and in my mood and speech'. The elusive metrical patterning echoes this mazy moonlit pattern directing the meditation, until the pattern unknits '[b]ack to its normal figure' in the last lines. The last two lines are couplets which take us out of the intervolved rhymes of the preceding stanzas:

> Said I then, sunk in tone,
> 'I am merest mimicker and counterfeit!—
> Though thinking, *I am I*,
> *And what I do I do myself alone.*'
> —The cynic twist of the page thereat unknit
> Back to its normal figure, having wrought its purport wry,
> The Mage's mirror left the window-square,
> And the stained moon and drift retook their places there.

From 'A Cathedral Facade at Midnight' to 'The Pedigree', Hardy achieves a fascinating series of metrical structures of unique expressiveness. One of the most remarkable aspect of Hardy's maturing use of versification is his use of the visual stanza as a mimetic tool. To this significant contribution to the history of metrical practice and theory, we shall turn in Chapter 5.

The series we have traced are highly selective ways of isolating Hardy's metrical effects. Of course, there are many more wide ranging and flexible ways to treat Hardy's metres. Here we can only nod briefly toward 'After A Journey', described above as Hardy's most metrically interesting poem. The poem also comes towards the end of a great series of journey rhythm poems: 'My Cicely', 'In Tenebris ii', 'Wessex Heights', 'Valenciennes', 'The Colonel's Soliloquy', 'The Going of the Battery', 'The Supplanter', 'A Trampwoman's Tragedy', 'The Night of Trafalgár', 'Budmouth Dears', 'Beyond the Last Lamp', 'St. Launce's Revisited', 'A New Year's Eve in War Time', ' "Quid Hic Agis" ', 'The Five Students'. The poems in this series represent Hardy's striking development of the English tradition of travel poem. They set up a strong rhythm which is associated with the rhythm of journeying, and which subtly (and not so subtly) moulds the consciousness of the speaker. An ancient source of this mimesis is Horace's fifth satire (Book i): 'Brundisium ends both my long journey, and my paper' (in Smart's translation in the edition owned and much marked by Hardy).

But here I mean only to allude to a wider-ranging way of discussing Hardy metres—consistent with Coleridge's point that 'variation . . . is not introduced wantonly, or for the mere ends of convenience, but in correspondence with some transition in the nature of the imagery or passion'.

We have seen the dipody in 'After a Journey'. Each line shimmers with expressive possibilities. The competing pentameter and tetrameter norms often produce dramatic shifts in emphasis in the manner of 'Near Lanivet'. While the tetrameter lilt seems to entrance Hardy, the pentameter possibility seems to interrupt the lilt and emphasize other things.

For example it emphasizes the lovers' personal relationship in all its painful continuing reality. (The stress marks I give picture the basic tetrameter lilt; below I have underlined those candidates for the dipodic potential accent urging the pentameter rhythm):

Thróugh the years, thróugh the dead scénes I have trácked you

I sée what you are dóing: you are léading me ón

And nót the thin ghóst that I nów frailly fóllow

Sóon you will háve, Dear, to vánish from mé

That it séems to call óut to me from fórty years agó

Poignant personal reflectiveness continually intrudes on the lilt
of the old metre. The pentameter possibility also emphasizes
the here and now of the scene which intrudes so painfully on
the reverie:

> To the spóts we knéw when we háunted <u>here</u> togéther
> Ígnorant of whát there is flítting <u>here</u> to sée
> The brínging me hére; nay bríng me <u>here</u> agáin

And perhaps, if we alter an earlier scansion:

> <u>Hereto</u> I cóme to víew a vóiceless ghóst

Indeed sometimes the narrator seems to stumble about, and the
metre mimes this:

> Úp the <u>cliff,</u> dówn, till I'm lónely, lóst
> Fácing <u>round</u> abóut me éverywhére

Sometimes the uncertainty of the dipodic 'blank' accent creates
a hovering tentative effect, almost hectic and anxious, which
matches the meaning. In the following lines, the pentameter
possibility may make one revise the tetrameter scansion in the
direction of more uncertainty:

> <u>Where you will next be</u> there's no knowing
> <u>And gray eyes, and rose-flush</u> coming and going

Leavis has noted the 'vivid precision' of 'ejaculations' (which
we saw revised from 'soliloquies') in line 4, which Leavis
describes as conveying 'the slap of the waves . . . with its
prolonging reverberant syllables—the hollow voice, in fact,
that, in stanza three, "seems to call out to me from forty years
ago"' (129). The reverberation is helped by the slight dipodic
drag:

> And the únseen wáters ejáculations áwe me

With such a supple metrical instrument, Hardy conveys how
his reflections mingle with several kinds of motions: the ghost-
led journey, the ejaculations of stream and waterfall, the
beating of hearts together, the rhythms of morning life which

contrast with the evanescing ghost vision. Such 'onomatopoeia' is kept delicately subordinate to the movement of a mind, placating, arguing, grieving, holding out for the little that is left. The speaking voice is at once influenced by these onomatopoeic rhythms and resists them as well. For they are either the ghostly relics of what is no more or they issue from a natural world which buries the people we love.

Note: The Dynasts

Hardy's verse drama, *The Dynasts*, may be seen to occupy a kind of mid-point of Hardy's metrical development. In his stage directions to *The Dynasts*, Hardy will typically assert that men's voices fade away into 'the babblings of the sea' or into the 'sputtering' of the green wood fire (pp. 32, 343). We have seen how two series of sound symbolism poems develop these effects far beyond what the stage directions suggest.

The Dynasts plays an important role, in more essential respects, in Hardy's metrical development. The work was published in three parts, in 1903, 1905, and 1907. Its last part overlapped with the publication of Saintsbury's three-volume *History of English Prosody* in 1906, 1908, and 1910. Hardy took great pride in the drama's metrical aspects. As Gosse wrote to Hardy in 1908, 'Surely you must have been giving a great deal of attention to the technique of prosody? The ingenious turns of versification in this latest volume were not lost on me' (Gosse, *Life*, 309–10). Ford Madox Ford said Hardy talked by the hour about *The Dynasts*, 'explaining why he had used here heroics, here Alcaics or Sapphics or ballad forms or forms invented by himself . . . and keenly delighting in his achievement' (448). Less interesting metrically than the poems written afterwards, *The Dynasts* nevertheless pointed the way for (*a*) more complex combinations of the constituents of Hardy's verse, (*b*) new kinds of mind–nature mimesis, and (*c*) a growing awareness of the possibility of seeing metrical form as the final perfection and ossification of human speech.

Technically, for example, of the 244 poems in Hardy's first three volumes of poems, only eight poems (excluding sonnets) use stanzas longer than eight lines. Of these early poems, 40 per cent (98) are entirely in simple quatrains. Of the poems published after *Time's Laughingstocks* (Hardy's third volume, which includes poems written throughout the decade of *The Dynasts*), only 27 per cent (184 out of 672) are in quatrain form, and there are a hundred poems with stanza forms composed of nine or more lines. The longer stanza enabled Hardy to combine more of his verse constituents and achieve some of his greatest effects. *The Dynasts* is pivotal in this development toward the longer stanza form. Ten of its stanza forms are nine or more lines in length. *The Dynasts* also contains a total of forty-six new stanza forms. Only six of the lyrics in *The Dynasts* use as many as three constituents, but Hardy's willingness to combine more constituents increases noticeably thereafter.

More important than these tentative technical steps 'forward' are the suggestions about the expressive dimensions of metrical form which Hardy explores. *The Dynasts* is composed in three basic forms, the natural speech rhythms and dialect of everyday people, the blank verse and stately idiom used most commonly for parliamentary debate and military councils, and the rhyming verse stanzas spoken mainly by the Spirits. The Spirits, Hardy says, are the 'flower of Man's intelligence' (p. 137), 'the best human intelligence of their time in a sort of quintessential form' (*Collected Letters*, ii. 117); and they are contrasted therefore to the mass of unknowing men. The great generals and leaders occasionally see what the Spirits see. Thus there is a hierarchy of languages, matching the hierarchy of consciousness. As 'voices' become more knowing and self-conscious, they become more formal and metrical. Hardy often noted that lines, seemingly unpoetic out of context, are redeemed in the light of the entire poem. I am arguing here for one way of seeing the truth in that remark. If the Spirit's voices are more 'poetic' than the people's, there is an irony in this hierarchy, since the Spirit voices are ultimately ignorant of the Will which controls all things. The Spirits may 'muse' on the Will from whose 'stress' they 'emanate', but they can never 'learn' its laws or understand its nature. The intelligence they express is thus a fruitless intelligence, whose apparent insights are cancelled by unforeseeable changes in the Will. The forms of expression of this intelligence are thus momentary and delusory 'abstractions' from the Will's ongoing reality. Thus the 'Spirit of the Years' addresses the 'Chorus of Pities', and gives us one more version of Hardy's Aeolian lyre:

> Why must ye echo as mechanic mimes
> These mortal minions' bootless cadences,
> Played on the stops of their anatomy
> As is the mewling music on the strings
> Of yonder ship-masts by the unweeting wind,
> Or the frail tune upon this withering sedge
> That holds its papery blades against the gale?[21]

The Spirit of the Years refers not only to the mechanical cadences whose implications we have discussed, but also to the mingling of such cadences with the sounds of wind and storm. What *The Dynasts* (and the novels) assert, the poems dramatize with increasing skill. In them we seem to hear in the metrical form the blending of voice and weather into a final mechanic echo.

---

[21] *Dynasts*, pp. 137, 7, 252. Also note p. 212: Napoleon, in a 'spiritual' mood, 'sinks into the rigidity of profound thought'. Hardy tended to compose *The Dynasts* by blocking out the verse sections in prose and then converting them to verse (Millgate, *Biography*, 452): thus, his process of composition matched the evolution of metricality which he dramatized.

5. Hardy's Shaped Stanzas and the Stanzaic Tradition

As Hardy creates more original stanza forms for more complex mimetic effects, the visual form of the stanza assumes more and more importance. A Hardy impression, the interaction of mind with movements in the outer world, culminates in a verse shape, which is the final stage of the impression. This shape becomes a visual icon of the poem's meaning. The visuality of the Hardy stanza has an interesting relationship to the tradition of inscription poetry, the history of accentual syllabic theory, and the modern transition to free verse.

Hardy and the Tradition of Inscription Poetry

Hardy's poetry is often reminiscent of the ancient tradition of visual inscription poetry, epitaphs, and graveyard carvings. Many of Hardy's characteristic poems are written as epitaphs: 'Cardinal Bembo's Epitaph on Raphael' (i. 224), 'A Gentleman's Epitaph on Himself and a Lady, Who were Buried Together' (ii. 350), 'Epitaph' (ii. 481), presumably Hardy's own, 'Cynic's Epitaph' (iii. 115), 'Epitaph on a Pessimist' (iii. 124), 'A Necessitarian's Epitaph' (iii. 228), 'A Placid Man's Epitaph' (iii. 243), his two death-bed Epitaphs on Chesterton and George Moore (iii. 308–9), and many other poems that do not have 'Epitaph' in the title. Many Hardy poems present themselves as objects speaking, as in 'Inscriptions for a Peal of Eight Bells' (iii. 122), 'Haunting Fingers' (ii. 357), 'The Sundial on a Wet Day' (iii. 131): 'I drip, drip here / In Atlantic rain'; these link up with the tradition of inscription poetry where 'the inscription was anything conscious of the place on which it was written' (Hartman, 390). Other Hardy poems might be considered speaking epitaphs, the speech of the buried dead, as in 'Intra Sepulchrum' (ii. 469) and 'In Death Divided' (ii. 27) where the poem can be seen as the visualized form of 'the eternal tie' which 'No eye will [otherwise] see'. These poems are interesting versions of the tradition of lapidary inscriptions left on tombs and objects which are supposed to speak. Other poems are the voices of ghosts reduced to the natural objects of their graveyards, as in 'Voices from Things Growing in a Churchyard'. Other poems seem to be the speeches of gravestones themselves as in 'A Merrymaking in Question' (ii, 203). In other poems, like 'During Wind and Rain', the consciousness of the narrator seems, by the end of the poem, to metamorphose into the speech of the graveyard. A good example of this transmutation is 'Lying Awake', where the 'I' of the indented lines of the poem changes to the gravestone names in the final indented line:

You, Meadow, are white with your counterpane cover of dew,
    I see it as if I were there;
You, Churchyard, are lightening faint from the shade of the yew,
    The names creeping out everywhere.   (iii. 198)

This poem is Hardy's most deeply imagined version of the classical inscriptions where tombstones speak about themselves. The poem was published in Hardy's last and posthumous volume, appropriately entitled *Winter Words*.

This metamorphosis of 'I' speaking into 'it' speaking is a subliminal fact in many Hardy poems, as for example 'Logs on the Hearth' (ii. 232). As we have seen, the poem mimes the motions of a fire-led meditation. The speaker gazes at the fire, and his meditation joins with the burning log which itself seems to utter forth the spirit of his dead sister. This kind of mimesis, which we have traced in the last two chapters, points to this final sense of a transformation into an inscribed memorial, all that is left of the speaker and the scene. A similar fusion of mind and ghosts occurs in 'The Ghost of the Past', and the conclusion points to the time when what the speaker sees now will be fixed

into a dimming far-off skeleton. In effect, the verse skeleton of
the poem remains as the speaker's life fades, as an ironically
reverse ratio of what he asserts:

> And so with time my vision less,
>   Yea, less and less
> Makes of that Past my housemistress,
>   It dwindles in my eye;
> It looms a far-off skeleton
>   And not a comrade nigh,
> A fitful far-off skeleton
>   Dimming as days draw by.   (ii. 13)

Another example of this effect is 'His Immortality' (i. 180). As
in 'The Phantom Horsewoman' (ii. 65), the stanza form is the
final 'phantom of his own figuring', what might be called, in a
visual-aural analogy, a 'radiant hum' ('Voices from Things
Growing in a Churchyard', ii. 395). All that is left to the ear and
the eye is a verse skeleton.

' "Sacred to the Memory" ' (ii. 542) specifically extends the
notion of a tombstone inscription to the narrator's consciousness: 'my full script is not confined / To that stone space, but
stands deep lined / Upon the landscape.' Hardy sees his own
mind as a set of inscriptions: 'Enchased and lettered as a tomb,
/ And scored with prints of perished hands, / And chronicled
with dates of doom' ('On an Invitation to the United States', i.
142). His mind, he says in 'In a Former Resort after Many
Years' (iii. 8), is 'scored with necrologic scrawls, / where feeble
voices rise, once full-defined, / From underground in curious
calls'. He describes a heart 'inscribed like a terrestrial sphere /
With quaint vermiculations close and clear— / His graving'
('His Heart', ii. 198). 'The Inscription' (ii. 460) is a melo-
dramatic story version of this theme. A tomb inscription, with
the widow's death date left uninscribed, keeps her from
remarrying. The inscription thus imprints her consciousness, as
the poem imprints the reader's consciousness, a parallel Hardy
reinforces when he makes the last stanza repeat an earlier one,
and links widow and reader with the community repeating
ritual forms:

And hence, as indited above, you may read even now
The quaint Church-text, with the date of her death left bare,
In the aged Estminster aisle, where folk yet bow
        Themselves in prayer.

The role of the poem's visual shape in this imprinting is
emphasized by Hardy when he quotes the inscription in an old
English type style. On the tomb, the widow 'implored good
people to pray,

> '**Of their Charytie | For these twaine Soules**',

When Hardy sees his poems as recorded 'impressions', he is
conscious of them as inscribed impressions, impressions of a
world long gone. An impression is a one-time conjunction of
mind and world; it is by nature 'fugitive', and can only be
preserved in an inscription. In 'To My Father's Violin' (ii.
186), his empathic imagination joins with the object, and the
poem is the final epigraphic form of that union:

                                I sadly con
        Your present dumbness, shape your olden story.

The 'shape' Hardy alludes to is specifically and literally the
shape of the verse stanza. 'The Monument-maker' (iii. 14) is
also about a stanza-maker, and reminds us how deeply the
patterns of Gothic architecture are connected with Hardy's
stanza design:

        I chiselled her monument
            To my mind's content,
        Took it to the church by night,
        When her planet was at its height,
    And set it where I had figured the place in the daytime.
            Having niched it there
        I stepped back, cheered, and thought its outlines fair,
            And its marbles rare.

The irony of the poem, that the monument has little to do with
the real woman, reflects the fact that the monument is in the
speaker, a phantom of his own figuring, and that the poem is
his chiselling upon a blank page.

The poem as surviving inscription is Hardy's oddly literal

notion of the ancient idea (as in Horace's ode, 'Exegi Monumentum' III. 30) of the poem as immortal monument Hardy knew the tradition well in Shakespeare's sonnets (18, 60, 65 in Palgrave, also 55, 63), which also emphasize the poem as a seen thing: 'when you entombed in men's eyes shall lie' (81). Hardy also may have known Herrick's 'His Poetrie His Pillar' ('Behold this living stone, / I reare for me') and Spenser's sonnet which plays with the conceit: 'One day I wrote her name upon the strand, / But came the waves and washéd it away'. Hardy's 'Her Temple' (ii. 401) expresses the irony traditionally accompanying the encomium:

> Dear, think not that they will forget you:
> —If craftsmanly art should be mine
> I will build up a temple, and set you
> Therein as its shrine.
>
> They may say: 'Why a woman such honour?'
> —Be told, 'O, so sweet was her fame,
> That a man heaped this splendour upon her;
> None now knows his name.'

'Nor hers', it might be added. The visuality of Hardy's printed stanzas projects us back to the primitive motive for inscriptions, to preserve what is uttered. Yet an inscription is instantly old. Like his 'marble tablet' (ii. 433), Hardy's poem 'Gives all that it can, tersely lined'. The complex lines of 'Thoughts of Phena' (i. 81) embody a final embroidered design, which is asserted to be the relic of what is no more, a monument of verse structure built over an absence:

> Thus I do but the phantom retain
> Of the maiden of yore
> As my relic; yet haply the best of her—fined in my brain
> It may be the more
> That no line of her writing have I,
> Nor a thread of her hair,
> No mark of her late time as dame in her dwelling, whereby
> I may picture her there.

This pervasive visuality in Hardy helps explain the power of 'The Walk' (ii. 49). The second stanza repeats the form of the first stanza ('You did not walk with me . . .') but with an added self-consciousness, the difference of surveying the familiar ground, that is, seeing the familiar form as a remnant impression. Thus the stanza form is like the familiar room to which Hardy returns:

> I walked up there to-day
> Just in the former way:
> Surveyed around
> The familiar ground
> By myself again:
> What difference, then?
> Only that underlying sense
> Of the look of a room on returning thence.

The stanza-formed language is itself the 'difference' between the experiences recorded in the first and second stanzas. The stanza wears the sense of being looked up, for a sign of what remains.

In the article cited above, 'Wordsworth, Inscriptions, and Romantic Nature Poetry', Geoffrey Hartman notes that the 'votive inscription is important for nature poetry in that it allows landscape to speak directly' (394); Wordsworth 'transformed the inscription into an independent nature-poem', by drawing 'the landscape evocatively into the poetry itself' and 'by incorporating in addition to a particular scene the very process of inscribing or interpreting it' (400). 'Lines Composed a Few Miles Above Tintern Abbey', whose title reflects the inscription tradition, eventually transcends dependence on the setting, and makes the setting subordinate to a meditative consciousness proposed as a kind of permanent life, as though language and life could be unfadingly 'coterminous' (Hartman, 405).[1] The difference here between Wordsworth and Hardy is that Hardy's inscription poems are kept tied to their objects, real objects that existed and have now decayed, leaving only the husk of their inscriptions, the versified impression. In a sense Hardy's ultimate object speaking is the world, and the

---

[1] For a summary of the paradoxes shadowing the ideal of coterminous language and life, see my 'Natural Supernaturalism's New Clothes', *Wordsworth Circle*, 5 (1974), 33–40.

ultimate surviving inscription is not a voice or a person or
anything heard, but a series of marks, the visual writing.
Hardy's own critical theory of epitaphs can be glimpsed in an
interesting remark he made in a letter to Swinburne: 'examining several English imitations of a well-known fragment of
Sappho. . . . I then stumbled upon your "Thee, too, the years
shall cover", and all my spirit for poetic pains died out of me.
Those few words present, I think, the finest *drama* of Death and
Oblivion, so to speak, in our tongue' (*Life*, 287). In 'Sapphic
Fragment' (i. 222), Hardy himself imitated Sappho's lines,
which he found, along with Swinburne's translation (from
'Anactoria'), in his 1895 copy of Wharton's *Sappho*; he had
earlier marked the lines (189–95) in his 1873 copy of
Swinburne's *Poems and Ballads*, and next to them written 'From
Sappho'. The Sapphic epitaph occurred to him also when he
marked Shakespeare's line, where Henry resolves to conquer
France, 'Or lay these bones in an unworthy urn, / *Tombless,
with no remembrance over them*' (*Henry V*, i. ii. 229: see Wright, 14):
in the margin of this 1856 edition, Hardy then wrote 'cf
Sappho' and eventually used the line as one of the epitaphs for
his own poem, published in 1901. Hardy's version reads:

> Dead shalt thou lie; and nought
> Be told of thee or thought,
> For thou hast plucked not of the Muses' tree:
> And even in Hades' halls
> Amidst thy fellow-thralls
> No friendly shade thy shade shall company!

The '*drama* of death and oblivion' which Hardy sees is the
drama implicit in a visual graveyard inscription. The voice of
someone long extinguished speaks out of the grave to someone
soon to be extinguished, with only the epigram surviving in
some physical form.

A great normative text, I suspect, for Hardy's conception of
his poetry is the *Greek Anthology*. He owned a 1911 edition of
Select Epigrams from the Greek Anthology, edited by J. W. Mackail,
but presumably knew the anthology much earlier as part of his
classical education. His reading of Mackail's selection lead to

some conscious imitations: 'The Bad Example' (iii. 171) drawn from Mackail 8.3, 'Faithful Wilson' (iii. 233) from Mackail 1.76, and 'Epitaph on a Pessimist' from Mackail 3.65. Mackail's introduction (especially pp. 3, 11, 27–8) discusses the problem of defining and categorizing the epigrams, much in the way Hardy discusses the problem of classifying poems which blend narrative, lyrical, and meditative (above, Chapter 2). The *Anthology*, evoking the root meaning of the words, brings together epitaphs, epigraphs, and epigrams. Its one unifying conception seems to be that of the poem as a brief text, a seen inscription, as in Mackail's definition: 'In brief then, the epigram in its first intention may be described as a very short poem summing up as though in a memorial inscription what it is desired to make permanently memorable in a single action or situation' (p. 4). Thus one of Sappho's epigrams proclaims: 'the white leaves of the dear ode of Sappho remain yet' (4.5).

Several of Hardy's poems play with the notion of the poem's marks being themselves shaped by natural forces, so that nature co-operates in creating the inscription. The poem evolves out of the world substance whose forces condition it. An analogy is given in 'The Abbey Mason' (ii. 124) where the architect's drawing is completed by the rain, which results in a coherent aesthetic form, the creation of the 'Perpendicular' style:

> The chalk-scratched draught-board faced the rain,
>
> Whose icicled drops deformed the lines
> Innumerous of his lame designs,
>
> So that they streamed in small white threads
> From the upper segments to the heads
>
> Of arcs below, uniting them
> Each by a stalactitic stem.

Thus the architect finds his structure. Like the architecture, so also the verse form. This connection is vividly demonstrated in 'The Figure in the Scene' (ii. 216), which adds an interesting mimetic dimension to the complex design of a poem like 'Thoughts of Phena'. Hardy said the poem was also 'possibly among the best I have written' (*Collected Letters*, vi. 96).

It pleased her to step in front and sit
    Where the cragged slope was green,
While I stood back that I might pencil it
    With her amid the scene.
        Till it gloomed and rained;
But I kept on, despite the drifting wet
    That fell and stained
My draught, leaving for curious quizzings yet
    The blots engrained.

And thus I drew her there alone,
    Seated amid the gauze
Of moisture, hooded, only her outline shown,
    With rainfall marked across.
        —Soon passed our stay;
Yet her rainy form is the Genius still of the spot,
    Immutable, yea,
Though the place now knows her no more, and has known her not
    Ever since that day.

Here and in 'Copying Architecture in an Old Minster' (ii. 171), Hardy plays on the analogy of the sketching pencil and the poet's pen, the sketching pad and the poet's paper. Here the drawing is marked by the rain which leaves only an outline of the woman, an outline which acts like an inscription carved by man and confirmed by nature. The point to be emphasized is that the poem's verse form is a similar inscription, its jagged wavery form miming the interaction of writing and rain. We are left with this mimesis, but a mimesis now frozen in verse. The figure in the scene has become the poem in the eye.

The conjunction of Hardy's epitaphic sense of the poem's stanza form, its survival as a mimesis conditioned by nature, and its status as a final epigraph of a poet long since departed, can be seen in one of Hardy's masterpieces, 'At Castle Boterel'. Its themes, echoed in metre, include journeying, rainy weather, the flickering and diminishing of the light of vision. Its form is merely common metre with an added dimeter line, the same form used in 'Old Furniture', a late poem in the firelight series, and also similar to the form of Swinburne's 'Félise', which is thematically related. Hardy varies the rhythm and makes the visual shape of the stanza contribute to his theme:

As I drive to the junction of lane and highway,
   And the drizzle bedrenches the waggonette,
I look behind at the fading byway,
   And see on its slope, now glistening wet,
      Distinctly yet   (ii. 63)

The poem poignantly exaggerates the literal seeing of an old image: 'Primaeval rocks . . . record in colour and cast . . . that we two passed.' The rhythmic variations beautifully capture the complex interaction of two journeys, the one the speaker now takes in the rain, the earlier one he took with his lover on a spring day. Their point of intersection occurs in the vision of a 'phantom figure' which fades in the gloom and shrinks in the rain:

     I look and see it there, shrinking, shrinking,
       I look back at it amid the rain
    For the very last time; for my sand is sinking,
      And I shall traverse old love's domain
         Never again.

The stanzaic pattern helps outline that figure both in its belatedly distinct and presently eroding quality. This pattern seems to capture the way the rain outlines a figure in a brief watery outline and the way light focuses a figure in a fading frame. Hardy's meditative voice interacts beautifully with this sound symbolism and the stanza's visuality.

What we have seen then is Hardy's interesting version of the Aeolian lyre myth. And here we can see several aspects of our theme come together: sound symbolism in the Aeolian lyre tradition, the figure of echo, and the poem as epitaph. In 'After the Fair', for example, the Aeolian music of the chime is felt reverberating through the verse form; those chimes become an 'echo' extending through space and history and are finally preserved in the form of the poem—like a 'feeble voice . . . once full defined' in 'In Front of the Landscape' (ii. 7), and now become like the carved names at the end, appropriately, of 'During Wind and Rain': the consciousness moulded by nature ends in these last visual marks of intelligibility, which are defined by the same process that effaces them. The coalescence

of such traditions explains the odd association of the last line of
'During Wind and Rain' ('Down their carved names the rain-
drop ploughs') with the conclusion of Marvell's pastoral lyric,
'The Nymph Complaining for the Death of Her Faun'; 'my
unhappy Statue shall / Be cut in Marble. . . .'

> . . . I shall weep though I be Stone:
> Until my Tears, still dropping, wear
> My breast, themselves engraving there.

The speaker, her own Aeolian muse, weeps her own body lines
into a stony monument.

What lead Hardy to his constant experimentation with
stanza form? One answer is that he desired to preserve forever a
trace of the human; the ultimate horror for Hardy was that of a
'levelled churchyard' (i. 196), an 'obliterate tomb' (ii. 101). He
yearned to see consciousness made visible, to see, in a literal
sense, 'Moments of Vision', some trace of a consciousness
that is not eventually destroyed. 'Not God nor Demon
can . . . Unsight the seen', he said in 'To Meet, or Otherwise'
(ii. 15). He characterized himself as a 'A Sign-Seeker' (i. 65)
who spends his life looking in a very literal sense: 'I mark . . . I
see . . . I view . . . I have seen . . . I witness'. But he wishes to
see something else: 'those sights of which old prophets
tell . . . To glimpse a phantom parent, friend . . . Or, if a
dead Love's lips . . . Should leave some print . . . If some
Recorder . . . should . . . drop one plume':

> I have lain in dead men's beds, have walked
> The tombs of those with whom I had talked,
> Called many a gone and goodly one to shape a sign.

'I would cheerfully have given ten years of my life to see a
ghost—an authentic, indubitable spectre. . . . I am cut out by
nature for a ghost-seer' (Taylor, *Hardy's Poetry*, 106). 'I see the
ghost of a perished day', he writes ('A Procession of Dead
Days', ii. 420). 'I see forms of old time talking' ('The House of
Hospitalities', i. 255). In 'The Absolute Explains' (iii. 68) he
discovers a place in which past years are preserved like
inscribed parchment:

> There were my ever memorable
> Glad days of pilgrimage,
> Coiled like a precious parchment fell,
> Illuminated page by page,
> Unhurt by age.

In his signed copy of Cowley's *Works* (London, 1710) (Dorset County Museum), Hardy marked 'Ode: Mr. Cowley's Book Presenting Itself to the University Library of Oxford' (a source for *Jude the Obscure*), and underlined in the following lines: 'Where still the *shapes of parted Souls* abide / Embalm'd in verse.' *The Dynasts* is an attempt to picture world consciousness as the patterns of the Immanent Will. Here he said he attempted to render 'as visible essences, spectres, etc., the abstract thoughts of the analytic school' (*Life*, 177). He longed in art to create 'visible essences'. Yet his poems are full of absence, the missing. Seeing in Hardy is seeing too late. What he leaves us with is the stanzaic trace of a visualized consciousness now lost.

Visuality in the Accentual-Syllabic Tradition

The visuality of Hardy's poetry has an important relation to the accentual-syllabic tradition, and to Victorian metrical theory. As we have seen, the great achievement of the Victorian prosodists was their discovery of the abstract nature of metrical form. It had taken centuries to conceptualize this abstract form, and it had been traditionally tempting to reduce it to more quantifiable units. But the Victorians discovered that these quantifiable units, like syllables and accents, can vary, and yet the form remain. 'The Figure in the Scene' (cited above) works like a traditional accentual-syllabic poem. The stanzaic shape is preserved in the second stanza and so alerts us that the parallel lines of each stanza have the same metre, even though the number of syllables, accents, and accent placement in each line are very different. For example, the penultimate lines are pentameters, despite their great variety:

My dráught, leáving for cúrious quízzings yét
Though the pláce now knóws her no móre, and has knówn her nót

As we have seen, Victorian metrical theory suggests we often have to read the stanza in its entirety, or even read the subsequent stanzas, before we can securely determine the poem's metre. The stanza is in fact a kind of picture of the abstract metrical form which defies quantification. The great analogy, again, is Gothic architecture. In his 1906 essay, 'Memories of Church Restoration', Hardy describes Gothic architecture in terms that can easily be adapted to the Victorian theory of prosody:

It is easy to show that the essence and soul of an architectural monument does not lie in the particular blocks of stone or timber that compose it, but in the mere forms to which those materials have been shaped. We discern in a moment that it is in the boundary of a solid— its insubstantial superfices or mould—and not in the solid itself, that its right lies to exist as art. The whole quality of Gothic or other architecture—let it be a cathedral, a spire, a window, or what not— essentially attaches to this, and not to the substantial erection which it appears exclusively to consist in. Those limestones or sandstones have passed into its form; yet it is an idea independent of them—an aesthetic phantom without solidarity, which must just as suitably have chosen millions of other stones from the quarry whereon to display its beauties. Such perfect results of art as the aspect of Salisbury Cathedral from the north-east corner of the Close, the interior of Henry VII.'s Chapel at Westminster, the East Window of Merton Chapel, Oxford, would be no less perfect if at this moment, by the wand of some magician, other similar materials could be conjured into their shapes, and the old substance be made to vanish for ever.

This is, indeed, the actual process of organic nature herself, which is one continuous substitution. She is always discarding matter, while retaining the form. (*Personal Writings*, 213–14)

Accentual-syllabic metre is another such form, which persists through myriad variations of syllables and accents. Its essence does not lie in any particular combination of these, and they are subject to continuous substitution while the form is retained. The Victorian theory of accentual-syllabic form accorded well with Hardy's sense of the stanza form as an architectural monument in writing, 'an aesthetic phantom without solidarity'.

The dimension of visuality in Hardy's stanzas has interesting implications for the history of verse forms and verse theory. The theory of accentual-syllabic metre has long been dependent on various visual structures and analogies. Scanning of classical quantitative verse and of its true English imitations was a largely visual exercise; when Englishmen tried to hear classical feet, they usually heard and wrote accentual equivalents of classical quantities. The classical influence carried over into the use of 'feet' in English prosody; feet worked reasonably well as visualized divisions of the line, rather than as aural units. Indeed, the notion that feet are aural realities has caused much mischief in the history of metrical theory. The musical analogy, so important for one stage of metrical theory, was not entirely an aural analogy. It depended on the creation of a uniform system of bar description; thus the musical school 'saw' the verse line divided by bars, each of the boxes filled with a differing number of musical notes. The great Victorian insight, which Hardy copied from Patmore (above, Chapter 1) was that 'the sequence of vocal utterance . . . shall be divided into equal or proportionate spaces'. That these spaces were sometimes (and loosely) called 'isochronous' hid the degree of visuality in the way they are described. The submerged visuality in Victorian theory can be seen in the analogy of the 'skeleton rhythmus' of Thellwall, the 'verse skeleton' of Hardy, Ellis, Earle, Saintsbury, and Bridges (see above, Chapter 2).

In effect, our sense of English accentual-syllabic rhythm seems to result from a coalescence of visual and aural organization, resulting in a sense of spaced sound, sound falling into place, a space of time. John Hollander's book, *Vision and Resonance*, captures in its title the complementary aspects of our prosody; his first chapter is entitled 'The Poem in the Ear', his last 'The Poem in the Eye': 'The fact remains that from the early sixteenth century on, all poems are in some sense shaped' (268). We have seen that some theorists proposed that rhyme was necessary for rhythm, because rhyme defined the 'space' of a line. Once the idea of that space was established, then blank verse was possible. Johnson thus spoke more than he knew when he quoted a remark that Milton's blank verse 'seems to be

verse only to the eye'. When Donne referred to the 'pretty
rooms' of the stanza in 'The Canonization', and when
Wordsworth referred to the 'narrow room' of the sonnet, they
evoked the Italian etymology of the word, 'stanza'. As Dr
Johnson said in his dictionary definition: '*Stanza* is originally a
room of a house, and came to signify a subdivision of a poem.'
Mathematics and geometry, choreography (as in strophe and
antistrophe), architecture, inscriptions, the *ut pictura poesis*
tradition, the visible charting of madrigal and air patterns, and
the influence of printing (especially since Tottel's *Miscellany*
used indentation so influentially to indicate stanza structure)
were powerful influences on the construction of stanza forms.
Certainly their influence cannot be overestimated for Hardy,
whose aesthetic sensibility was formed by the Gothic architecture revival, the availability of painting in nineteenth-century
museums, and the revival of the *ut pictura poesis* tradition in Pre-
Raphaelite poetry.[2]

Paradoxically the visual architecture of the stanza has been
overlooked in the history of metrical theory. The *OED* defines
the stanza as 'a group of lines . . . arranged according to a
definite scheme which regulates the number of lines, the metre,
and (in rhymed poetry) the sequence of rhymes'. This definition reflects the way the word has been used, but omits an
important aspect of the 'definite scheme', the arrangement of
lines through indentation and margination. It has been very
hard for theorists to 'see' stanzas, to see them as the visualized

---

[2] Modern metrists have barely begun to speculate about visuality. John Hollander,
in 'The Poem in the Eye', says: 'The kinds of linguistic depth which these arrangements
of printed surface stir up are beyond my grasp and my subject' (271). Wimsatt's essay,
'In Search of Verbal Mimesis', ends with the speculation, 'we can ask whether the
visual does in fact join and accentuate a diagram present independently in the semiotic
and prosodic structure . . . or whether in fact . . . we have [technopaignia]'. Early
difficulties in sorting out the visual and aural nature of accentual-syllabic rhythm may
account for the oddities of Wyatt's metres. Scholarly post-McLuhan consideration of
the effect of print on culture and literature is reviewed in Walter Ong's *Orality and
Literacy* (London, 1982); most notable, perhaps is Elizabeth Eisenstein, *The Printing
Press as an Agent of Change*, 2 vols. (Cambridge, 1979). After Gutenberg, the preservation
of the text was no longer associated with carefully inscribed single manuscripts but with
multiple disposable copies. Thus, the stanza is a visualization of an entity no longer
identified with a material base but now floating free, with a kind of ghostly status
explored in Wellek and Warren's 'The Mode of Existence of a Work of Art'.

form of a heard rhythm which cannot be easily quantified. Indeed, the *ut pictura poesis* tradition ironically kept theorists blind to stanza shape, as they looked elsewhere for the picture in the speech. A great early exception to this visual blindness is George Puttenham, the first to use the word 'stanza' in its common spelling. In Book II, Chapter xi, 'Of Proportion by Situation', he outlined stanzas in the following way, to reveal rhyme pattern and line length. In Fig. 3 the drawing on the left (p. 92) indicates an *ababcc* rhyming pattern; the drawing on the right (p. 94) indicates a line-length pattern. Puttenham makes an observation that was eventually lost in the history of accentual-syllabic theory:

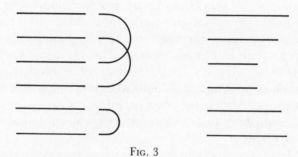

Fig. 3

And I set you downe an occular example, because ye may the better conceive it. Likewise it so falleth out most times your occular proportion doeth declare the nature of the audible; for if it please the eare well, the same represented by delineation to the view pleaseth the eye well, and *e converso*; and this is by a natural *simpathie* between the eare and the eye, and betweene tunes & colours, even as there is the like between the other sences and their objects, of which it apperteineth not here to speake. (89)

It is in the context of these visual diagrams that Puttenham uses an architectural analogy, comparing the 'bands' which hold together lines in a stanza to the bands which hold together the parts of a wall (93); similarly in his 1619 preface to his *Poems*, Drayton will use diagrams and compare his octave stanza to the Tuscan pillar in architecture.

Puttenham's insight came at the beginning of the great flourishing of the Elizabethan and seventeenth-century stanza, and was consistent with the matured versification of that age.[3] His insight was eventually lost as the age of Bysshe, in love with linear counting, grew distrustful of metaphysical wit and elaborate forms. (The convention of the ode, and the glamour of the musical analogy, kept the complex stanza alive as an exceptional form.) The Romantics recaptured the flexible line metric of the seventeenth century, but largely confined themselves to traditional stanzas. Indeed, Wordsworth and Coleridge never did devise an adequate justification for the artifice of metre. When the Victorian age developed its matured theory of accentual-syllabic metre, this triumph was accompanied by an extraordinary renewed flourishing of the stanza form. The importance of the stanza in metrical history is one reason, perhaps the only reason, why Saintsbury cannot disappear from our consideration. The greatest value of his work may be as a history of the stanza with examples and with discussion of how metrically effective the stanzas are. (Schipper, *Englische Metrik*, by contrast, simply schematizes stanza types.) Saintsbury's discussion is not very deep, but has never been superseded.

We have seen that when the Victorian period developed its insight into the abstract nature of the verse structure, it then had a way to talk about the mimetic possibilities of that structure. We have also seen that the Johnsonian objection to sound symbolism, 'representative metre', could only be answered by careful attention to the way sound served sense. This principle also applies to stanzaic mimesis, which waits upon the meaning which it elaborates. Interestingly, Saintsbury stumbled on the principle of iconic stanzaic form, but then withdrew his foot from the unfamiliar stream. Quoting the stanza of Browning's 'My Star', he says it illustrates 'the sovereign importance of the *line*' and challenges disbelievers 'to

---

[3] John Thompson has described this maturity in what might be called micro-metric terms, through attention to the line; but we might distinguish a macro-metric approach, attentive to the stanza form.

ask themselves why Browning divided the early lines and not the later'. He footnotes cryptically: 'There is, by the way, a prosodic and critical as well as an astronomical allegory here' (iii. 228–9). About 'James Lee's Wife', he notes that Browning 'never devised a better form for passionate *meditation* than the fretted outline of Part Two' (231). He then discusses what the stanzaic patterns of the other sections 'stand for' (232). He concludes: 'Indeed Browning has in few places better exempli-fied that higher suiting of sound to sense which is part of the nineteenth century's prosodic discoveries' (239). Hardy was at the forefront of this particular prosodic discovery. In 1925 Hardy would read in Henry Newbolt's anthology, *The Tide of Time in English Poetry*, principles which his own poetry had helped make explicit: 'In reading poetry we are frequently aware that every stanza-form has a mood-creating or thought-compelling power of its own, and helps very materially, if it is well chosen, to convey the sound of the meaning' (89). Hardy said the anthology 'has interested me much' (*Collected Letters*, vi. 326).

This great historical tradition of the shaped stanza must be distinguished from another slighter tradition, that of the figured poem, 'technopaignia', picture poetry, treated as 'pretie amourets' by Puttenham in his chapter, 'Of Proportion in Figure', which is distinct from and follows after the chapter, 'Of Proportion by Situation'. Viewed as 'madd gugawes and crockchettes' by Gabriel Harvey, and as 'false wit' by Addison, Dryden, Johnson, and their followers (including most new critics), technopaignia has also helped keep the English stanza from being properly seen. Several ancient examples, Simias's egg and wings and axe, Theocritus's pipe, Besantinus and Dosiadas' altar poems, can be seen in the *Greek Anthology*, xv. 21–7, though not in Mackail's selection. Also Hardy had read Herbert, Herrick, and Jonson, and so probably knew Herbert's picture poems ('The Altar', 'Easter Wings'), Herrick's 'The Pillar of Fame' (which concludes *Hesperides*) and 'This Crosse-Tree Here'. Ben Jonson's 'The Houre-Glass' is an intermediate example, while Herbert's 'The Collar' is more like the mimetic visuality we are exploring in Hardy. Interestingly, none of

Hardy's poems are properly technopaignic. 'The Peace Peal' (iii. 119) is perhaps the closest approximation:

> Said a wistful daw in Saint Peter's tower,
> High above Casterbridge slates and tiles,
> 'Why do the walls of my Gothic bower
> Shiver, and shrill out sounds for miles?
> 　　This gray old rubble
> 　　Has scorned such din
> 　　Since I knew trouble
> 　　And joy herein.
> 　　How still did abide them
> 　　These bells now swung,
> 　　While our nest beside them
> 　　Securely clung! . . .
> 　　It means some snare
> 　　For our feet or wings;
> 　　But I'll be ware
> 　　Of such baleful things!'
> And forth he flew from his louvred niche
> To take up life in a damp dark ditch.
> —So mortal motives are misread,
> And false designs attributed,
> In upper spheres of straws and sticks,
> Or lower, of pens and politics.

The poem takes on a tower like structure, but it stops short of being a literal picture of Saint Peter's tower. The structure echoes the meaning in more subtle ways, the shape also miming the narrow mentality of the bird and its final flight. The visuality of Hardy's poem is more suggestive than definitive, more flexibly waiting on meaning than imposing its picture. The same can be said of the major tradition of the lyric, as in the case, for example, of Vaughan's 'The Waterfall', which probably influenced the shape of Hardy's 'Under the Waterfall' (ii. 45). The fact is that technopaignia marks an extreme development of a pervasive aspect of stanzaic poetry, the contribution of visual form to its rhythm and its meaning. In technopaignia, the poem seems dominated by its visible form, which is recognizable by itself; indeed the printer must adjust the letter spacing to the picture. By contrast, in Hardy's poem, the visual form of the stanza has no meaning in itself and can only be properly seen when the poem is read. Technopaignia remains a kind of parody of genuine stanzaic iconicity. We might compare the way in which concrete poetry is an extreme pictorial form of a free-verse mimesis like Williams's 'Rain'. E. E. Cummings's play with the visual structure of letters is another extreme development of a pervasive fact of our verse. In mainstream mimetic verse like Hardy's, the stanza's visibility is an echo-icon of the thought.

Hardy and Free Verse

The visuality of Hardy's stanza forms has an important bearing on Hardy's relation to the free-verse tradition, and indeed throws an illuminating light on the perplexed relationship between the accentual-syllabic tradition and the free-verse tradition.

Ironically, when the English metrical theorists devised a theory adequate to the nature of accentual-syllabic verse, the accentual-syllabic tradition began to decline and be replaced by a free-verse tradition. It is a common notion that once the accentual-syllabic tradition had reached its final definition and its definitive practice in the poets, there was nothing left to do but turn to free verse. There is another possibility which would make free verse more consistent with the findings of the Victorian theorists. Once the sense of metrical abstraction had been formulated for accentual-syllabic verse by the Victorians, this same sense of abstraction became the basis for 'free' verse whereby any given line could take on the aura of abstract form. A similar phenomenon was happening in art, where notions of abstract form derived from traditional art were then pursued for their own sake. A *Poetry Journal* article of 1917 said: 'if a piece of writing is divided into "versicles" suggesting the arrangement of verse (whatever be its rhythmical character) we incline to accept it as a kind of verse product' (Leonard, 27). Bridges expressed this point somewhat scornfully in the essay 'Humdrum and Harum-Scarum' (the title referring to accentual-syllabic and free verse, respectively): 'However irregular the lines be, they are conscious of their length: they pose with a sort of independence and self-sufficiency.' Hardy copied from another portion of this essay, which he found 'excellent' (*Collected Letters*, vi. 165). In free verse, the sense of visual shaping, so influential in the history of the English lyric, becomes divorced from the standard accentual-syllabic stanzaic norm.

Paradoxically, then, the consciousness of abstract spacing, which finally made sense of traditional prosody, may have helped lead to that prosody's overthrow in favour of free verse, which separated the idea of metre from its dependence on syllables and accents. In free verse, the 'idea' of a number of syllables and the 'idea' of a number of accents valorizes whatever shows up in a given line. The abstraction, unearthed by the Victorians, could be separated from accentual-syllabic rhythm, the expected sequence of syllables and accents, and emerge in its own right. In much modern verse, traditional metrical rhythms may fade in and out like phantoms, or they may be entirely absent, but the verse as a whole retains the aura of metre. In free verse, we often have visual without rhythmic uniformity.

Interestingly, then, free verse has much to tell us about the meaning of traditional prosody, and traditional prosody has much to tell us about the meaning of free verse. Free verse has never known whether to justify itself as a visual or an aural phenomenon. Some of the earliest theorists emphasized its visual nature: 'This new verse . . . appeals to the eye rather than to the ear' (T. E. Hulme, 'Lecture on Modern Poetry'), though it was unclear whether this referred to images in the poem, or the poem as visual shape. Visual justification of free verse also fell athwart condemnations by critics who confused stanzaic iconicity with technopaignia, or who were influenced by the musical analogies of symbolist poets. Thus the more fashionable justifications of free verse have followed the language of traditional prosody, and have used aural concepts, like 'variable foot' (Williams), 'the sequence of the musical phrase' (Pound), 'cadence' (Amy Lowell), and so on; and yet the application of these concepts to variable line lengths, and to unsounded enjambments, has remained mysterious. The most obvious way to justify free verse—as visualized sound—has been slighted, because the visuality of traditional verse has never been properly conceived.[4]

In the 1860s, after he arrived in London, Hardy read Whitman (*Life*, 59) and indeed may have been among Whitman's earliest English readers. (The first English edition of Whitman's poems, edited by William Rossetti, came out in 1868.) What was Hardy's attitude to the growth of the free-verse tradition?

Hardy could not 'see' free verse. Neither he nor Hopkins were ready for such extreme literariness or for this next stage in the evolution of the metrical abstraction. Thus Hardy said that he saw free verse as a 'jumble of notes', 'the original sinner being Walt Whitman, who, I always think, wrote as he did, formlessly, because he could do no better'; 'only ancient poetry, like the English Bible, is able to dispense with [strict metre] because of its other character of antiquity' (*Collected Letters*, vi. 186, v. 192). Bridges's strictures on free verse in *Milton's Prosody* (1921) and 'Humdrum & Harum-Scarum' correspond well to Hardy's perceptions. He hoped Bridges's essay would 'tend to save our young poets from the woeful fogs of free verse worship' (*Collected Letters*, vi. 165). In the back of a volume of American poetry (Levy, 186), Hardy wrote:

The great charm of poetical form, as of music, lies in its relativity (to use the popular word), its relation to something received or expected. Free verse or free wording offers no attraction of that nature, except

---

[4] In the 1980s the visuality of William Carlos Williams's poetry has been explored in a number of essays and books, by Eleanor Berry, Marjorie Perloff, Henry Sayre, and, most recently, Stephen Cushman in *William Carlos Williams and the Meanings of Measure* (New Haven, 1985). These show the influence of Hollander's 1975 essay, 'The Poem in the Eye'. Recent interest in verse visuality, from concrete poetry to free verse, is also reflected in journals like *Visible Language* and *Word and Image*. On technopaignia in particular, see Jeremy Adler, '*Technopaignia, Carmina Figurata* and *Bilder-Reime*: Seventeenth-century Figured Poetry in Historical Perspective', *Comparative Criticism*, 4 (1982), 107–47. A fascinating story is reported by Robert Hass, that Robert Creeley learned his distinctive free-verse cadence by misreading some W. C. Williams poems and assuming that Williams's visual end-stops were sounded (Hass, *Twentieth Century Pleasures: Prose on Poetry*, (New York, 1984), 150–1).

perhaps that one sort of cadence leads you to expect another of the sort, which may or may not follow: it is demanded as in measure or rhyme.

Hardy found himself in the last period of English verse where accentual-syllabic expectation was the accepted norm. In the first version of his 1922 *Apology* prefixed to *Late Lyrics and Earlier*, he asks the liberty to 'quote Tennyson in this century of free verse' (*Complete Poems*, p. 562).

Nevertheless, Hardy is an important transitional figure between the accentual-syllabic and free-verse traditions. We need to look at some of Hardy's final experiments in verse.

Final Experiments

We have seen that Hardy's stanza forms are of different kinds and have varying functions. The stanzas can be (*a*) traditional forms, either repeated as in 'The Abbey Mason' or used once as in 'Epitaph on a Pessimist', (*b*) original forms, either repeated as in 'The Figure in the Scene' or used once. In the case of unrepeated original forms, a poem may use a number of these forms, as in 'The Pedigree', or may consist of one of these forms.

In the volumes following *Moments of Vision* Hardy publishes his most unusual experiments in stanza: the use of different stanza forms, mostly original, using three or more different line types, within the same poem. 'The Pedigree', with its four line types (3, 5, 6, and 7 beats) and each of its stanzas in a different form, is the beginning of this new series of metrical experiments. After 'The Pedigree' we see a marked incidence of poems which combine different stanzas where at least one of the stanzas uses three or more line types: there are two more such poems in *Moments of Vision* ('Fragment', ii. 261; 'In a Waiting-Room', ii. 266), two in *Late Lyrics and Earlier* ('At the Railway Station, Upton', ii. 377; ' "What did it mean?" ', ii. 431), eleven in *Human Shows* ('A Bird-scene at a Rural Dwelling', iii. 7; 'The Monument-maker', iii. 14; 'An East-End Curate', iii. 20; 'In St Paul's a While Ago', iii. 23; 'A Last Journey', iii. 25; 'A Spellbound Palace', iii. 28; 'One Who Married Above Him', iii. 49; ' "Not only I" ', iii. 101; 'The Sheep-boy', iii. 109; 'Shortening Days at the Homestead', iii. 133; 'A Hurried

Meeting', iii. 153). Seven of these (including four in *Human Shows*, iii. 25, 49, 101, 109) combine different stanzas in which at least one stanza uses four line types. The only examples of poems combining different stanza forms where at least one stanza uses five line types are the two in *Human Shows* ('Snow in the Suburbs', iii. 42; 'No Buyers', 48).

We also find that the *number* of different stanza forms in a given poem shows a dramatic increase after 'The Pedigree' (with its five different stanza forms). No poem before 'The Pedigree' uses more than two different stanza forms. Beginning with 'The Pedigree' in *Moments of Vision* there are eight poems from those just cited which use from three to five different stanza forms: two poems in *Moments of Vision* (ii. 261, 266), and six poems in *Human Shows* (iii. 28, 42, 48, 49, 101, 109). *Human Shows* also shows its remarkable character by containing the only two poems with six different stanza forms (iii. 25, 153).

Finally, we see an increase of single stanza poems in unusual forms. Previous single stanza poems by Hardy had been sonnets, triolets, couplets, or other familiar or at least uniform stanzas. But beginning with *Human Shows*, Hardy writes five poems in single stanzas with three or more line types: 'Ten Years Since' (iii. 31), 'Ice on the Highway' (iii. 45), 'The Sexton at Longpuddle' (iii. 95), 'The Mound' (iii. 176), 'Reluctant Confession' (iii. 179). 'The Sexton at Longpuddle' brings together our various themes, epitaph, visual stanza, tolling bell, and the transition to a more freely 'shaped' verse:

> He passes down the churchyard track
>   On his way to toll the bell;
> And stops, and looks at the graves around, ↴
> And notes each finished and greening mound
>   Complacently,
>   As their shaper he,
>   And one who can do it well,
> And, with a prosperous sense of his doing,
>   Thinks he'll not lack
> Plenty such work in the long ensuing
>   Futurity.
>   For people will always die,
>   And he will always be nigh
>   To shape their cell.

Such poems, with their unrepeated stanza forms, also impose
an unusual discipline on the reader, who has only one chance,
so to speak, to decide the metrical form of a given stanza.
Because lines and stanzas are not repeated, the structure comes
close to posing as free verse. Thus these poems constitute a kind
of missing link between the accentual-syllabic and free-verse
traditions. Nevertheless, the accentual-syllabic tradition keeps
the poems structured with some sense of expectant rhythm.
The way we discern this rhythm is through the visual stanzaic
indentation, the tetrameters unindented, the trimeters indented once, the dimeters indented twice.

'Ice on the Highway' (iii. 45) is another example that could
appear in a free-verse anthology:

> Seven buxom women abreast, and arm in arm,
> > Trudge down the hill, tip-toed,
> > > And breathing warm;
> They must perforce trudge thus, to keep upright
> > On the glassy ice-bound road,
> And they must get to market whether or no,
> > Provisions running low
> > With the nearing Saturday night,
> While the lumbering van wherein they mostly ride
> > > Can nowise go:
> Yet loud their laughter as they stagger and slide!

Again the visual indentation alerts us to the accentual-syllabic
tetrameters, trimeters, and dimeters; these same constituents
which in the previous poem helped build the shaped cell of the
sexton's epitaph, here are beautifully adapted to Hardy's
impression of the movement of these Brueghelesque women.

The development in Hardy's later career of more complicated stanza forms and his use of them for mimetic purposes
links him with parallel developments in poets as diverse as
George Herbert and William Carlos Williams. As his
accentual-syllabic career drew toward its close, it tended
naturally in the direction of the new poetry of his younger
colleagues. Some of these poems play with the transition
between traditional and freer forms. 'A Bird-scene at a Rural

Dwelling', for example, consists of two stanzas, the first a unique complex stanza, the second a conventional stanza of pentameter triplets. 'A Light Snow-fall After Frost' (iii. 43) plays with repeated and unrepeated stanzas: the first and last stanzas match, while the two middle stanzas are unique. We see a constant interplay of conventional and original, the repeated and the unique. 'Snow in the Suburbs' (iii. 42) is a major example of this late development in Hardy, with its two long complex original stanzas, and its concluding quatrain. Not until Hynes's Oxford edition in 1985 was the poem published in its correctly indented form, with the indentation consistently indicating the metrical line types. It is one of only two Hardy poems which use five line types in one stanza:

> Every branch big with it,
> Bent every twig with it;
> Every fork like a white web-boot;
> Every street and pavement mute:
> Some flakes have lost their way, and grope back upward, when
> Meeting those meandering down they turn and descend again.
> The palings are glued together like a wall,
> And there is no waft of wind with the fleecy fall.
>
> A sparrow enters the tree,
> Whereon immediately
> A snow-lump thrice his own slight size
> Descends on him and showers his head and eyes,
> And overturns him,
> And near inurns him,
> And lights on a nether twig, when its brush
> Starts off a volley of other lodging lumps with a rush.
>
> The steps are a blanched slope,
> Up which, with feeble hope,
> A black cat comes, wide-eyed and thin;
> And we take him in.

There is a big difference between 'And we take him in' in Hardy's accentual-syllabic rhythm, and the same line if it appeared in a William Carlos Williams poem; but there is a common tradition of the shaped poem co-operating with the meaning and rhythm of the language. This shape in the first stanza follows the static patterns of the snow, and in the second stanza the kinetic action of the bird. The final stanza is about seeing and being seen. 'We' have been looking on all this time; 'we' have been observing this scene and these shapes; 'we' have been seeing-saying the poem.

Afterword

A CONVENIENT place to read many of Hardy's reviews is in his own personal folder, entitled *Scrapbooks*, in the Dorset County Museum (see *Original Manuscripts*). In 1918 Hardy went through his reviews, threw out a few, but kept most. The reviews show a consistent pattern. Thirty-two of them attacked his clumsy and amateurish versification. 'Halting and jerky' (1899), 'his verse often halts or dances in hobnails' (1902); his 'slow muffled cadences . . . suggest old people dancing an old dance' (1909); 'the mechanism of a stanza creaks and groans with the pressure of its working' (1915). 'His verse often halts or dances in hobnails' (1920). Hardy noted that later reviews often quoted earlier reviews: 'All the American papers had picked up this phrase, and he had a number of cuttings— "Failure of Mr. Hardy's poems. No more music, etc., etc." ' What Hardy said of the American reviews was also true of the English. 'As critics have pointed out . . .', notes a *TLS* reviewer in 1917. In exasperation Hardy underlined the misspelling in the following sentence from a 1921 review in *Plain English*: 'Mr. Hardy is not even a moderately good *vesifier*' ('sic'). Gosse had been critical of Hardy's metrics in an 1899 review in *Literature*—though he had then offered the now familiar theory that Hardy was a 'signal example of the temporary success of a violent protest against the cultivation of form in verse'. However, by 1918 Gosse had changed his view. Consoling Hardy for his reviews, he wrote to him: 'Your form is abundant, excellent, and deserving of careful analysis. It is one of your strong points. Without 'form' poetry is void, and it is part of your genius as a poet that you are always so interesting linguistically and prosodically.' In 1923 Mark Van Doren referred to the 'time-old questions about Mr. Hardy's form, or lack of form'. After Hardy's death, his metrical reputation perhaps reached its low point in William Empson's observation in 1940: 'Hardy often simply drops his rhythm, as a child drops its rattle and stares before it straight at the skyline, dribbling slightly.' In 1942 Edmund Blunden referred to the 'still vexed questions of prosody and its diagram or category' in Hardy. The first full-length book on Hardy wearily concluded: 'He has made no real contribution to English prosody in spite of his experimentation.'[1]

Hardy was sensitive to such criticism: 'Can there be anything more paralyzing than to know that features, subjects, forms, & methods, adopted advisedly, will be set down to blundering, lack of information, pedantry, & the rest.' Repeatedly he insisted that his metrical techniques were 'cunning'. Whatever awkwardness he felt on returning to poetry in the late 1890s, he said, 'soon wore off'. He did not appreciate his reviewer's jokes: 'T.H. is only moderately pleased with the article', his wife Florence wrote, 'as he disagrees with Mr. Courtney's statement that "in answer to Nature's stern imperative 'Thou shalt not be a poet' he (T.H.) had boldly and laboriously answered 'I will' "'.' He complained of the critical tendency to 'hearken for

---

[1] On Hardy collecting reviews, see Millgate, *Biography*, 522. The quotations are in order from the *Pall Mall Gazette* (6 Jan. 1899); *Saturday Review* (11 Jan. 1902); the *Morning Post* (9 Dec. 1909); *The Bookman* (Feb. 1915); *Dial* (Jan. 1920); Felkin, 31; *TLS* (13 Dec. 1917); *Plain English* (26 Feb. 1921); *Literature* (4 Mar. 1899); Gosse, *Life*, 417; Van Doren, 'Lyrics and Magic', *Nation*, 116 (1923), 125; Empson, *New Statesman*, NS 20 (1940), 263–4; Blunden, 243; Southworth, 221. The other reviews citing Hardy's clumsy rhythms are: *Westminster Gazette* (11 Jan. 1899); *Daily Chronicle* (21 Dec. 1899); *Cosmopolitan* (Mar. 1899); *Academy* (14 Jan. 1899); *Saturday Review* (7 Jan. 1899) ('But as we read . . . these many, slovenly, slipshod, uncouth verses' etc.); *London Times* (5 Jan. 1899); *The Speaker* (24 Dec. 1898); *Literature* (28 Dec. 1901) ('Mr. Hardy's lines . . . have the air of being rough-hewn'); *Spectator* (5 Apr. 1902); *Dial* (1 May 1902); *Pall Mall Gazette* ('De Profundis' by G. S. S., *c.* 1902) ('Some of them read like pieces of prose turned forcibly into rhyme'); *Daily News* (13 Dec. 1909): 'The technical side of Mr. Hardy's poetry has been sufficiently illustrated' i.e. by reviews like these; *Country Life* (11 Dec. 1909); *Oxford Magazine* (24 Feb. 1910); *Pall Mall Gazette* (13 Jan. 1910); *Nation* (16 Jan. 1915); *Spectator* (2 Jan. 1915) (His rhythms represent 'a stubborn conquest'); *Chicago Evening Post* (2 Jan. 1915); *Nation* (4 Feb. 1915) ('By turns he is rough, dissonant, mean, bald, pompous, grotesque'); *Outlook* (22 Dec. 1917); *Saturday Review* (22 Dec. 1917); *Plain English* (26 Feb. 1921); *Dial* (June 1927); *Nation and Athenaeum* (13 Oct. 1928) ('Stiff, uncomfortable, rheumaticky'); *Observer* (21 Oct. 1928).

the key-creaks and be deaf to the diapason'. Till the day he
died, he looked for 'finely-touched spirits' attuned to the 'right
note-catching'.[2] To prove that he 'was not at this time and
later the apprentice at verse that he was supposed to be', Hardy
wrote the long assessment in the *Later Years* which we have
examined above. He was proud of his verse skeletons 'mostly
blank, and only designated by the usual marks for long and
short syllables, accentuations, etc., but ... occasionally made
up of "nonsense verses"'. But this defence only added fuel to
the fire. In answering one set of critics, he fell prey to another
more modern set of critics who complained that Hardy made
verse 'as if the elaboration of complicated forms were a virtue in
itself' (Howe, 881). 'It is possible ... to speak of Hardy's *choice
of metre* in a way that we would be reluctant to do in the case of
Hopkins, Eliot, Pound or Yeats, and, even more, to pass
judgement on that choice by designating it an arbitrary one'
(Hollander, *Vision*, 137). 'For Hardy, that is, the inseparability
of form and content, which we set so much store by, simply did
not exist' (Hynes, 76, after quoting Hollander). 'Hardy's
prosody often fails as an expressive form. ... Words are forced
into the metrical patterns; the metres themselves are frequently
inappropriate to the subject' (Gross, 42–3). 'The stanza
becomes a torture box' (Perkins, 147) constructed by 'the
imperious verbal engineer' (Davie, 59). The puzzle of Hardy's
failures has been that their list keeps changing: 'The charm in
rereading Hardy's poetry is that one keeps making discoveries'
(Brooks, 'Note', 69). 'Too many of Hardy's poems ... are not
"good". And I agree; but I am always changing my mind as to
which ones those are' (Van Doren, *Autobiography*, 167).

In the early reviews of Hardy's poetry, we also meet an
interesting riddle. At some point in many readers' experiences,

---

[2] Hardy, *Collected Letters*, ii. 300; *Life*, 301, 292; *Friends*, 296; *Complete Poems* pp. 560,
559. Southworth, 166, says Hardy's awkwardness may be caused by 'frequent attempts
at enclosing the thought in a pre-determined pattern, ill-fitted for the purpose'. Donald
Davie (17) blames such a fault on the influence of 'industrial technology in the age of
heavy engineering'. The complaint that Hardy's will does the work of his imagination
is heard occasionally in the early reviews: for example, Padraic Colum, 'Robert Bridges
and Thomas Hardy', *New Republic*, 12 (11 Aug. 1917), 47–9.

Hardy's verse forms assume a rich and interesting life. Hardy creates 'effects at times that penetrate and pervade our emotional consciousness with something of magic' (1909); 'he continues to perform miracles with a style which would at once sink any other poet to the bottom' (1918); 'A powerful music is made out of the most intractable material' (1922). If thirty-two of the early reviews find awkwardness and predetermined pattern in the rhythms, twenty-three reviews (including some of the above) in Hardy's collection find these defects transformed by a mysteriously compelling cadence or a rich musicality. One of the most intriguing testimonies to Hardy's magic is Siegfried Sassoon's: 'I suppose Hardy's lesser poems will always be appreciated only by a few people with the special palate for them, they are so obviously 'versified'; & yet they are unlike anyone else, & seem to acquire a greater intensity now that he is so far away from us, bless him.' Hardy himself claimed that when his poems were read aloud, they would sound 'not harsh, only inevitable'. He had copied down Arnold's extensive passage describing Wordsworth's style as 'inevitable as Nature herself'. In 1919 Middleton Murry, using Hardy's adjective, called Hardy's moments of vision 'inevitable'. The Hardy poem, he said, is not 'the record, but the culmination of an experience, and . . . the experience of which it is the culmination is far larger and more profound than the one which it seemed to record. . . . His reaction to an episode has behind and within it a reaction to the universe. . . . In it we may hear the sombre, ruthless rhythm of life itself.' Murry said that Hardy read these remarks and: 'I am happy to have his endorsement of the theory in a letter which he wrote to me.'[3]

---

[3] The quotations are in order from the *Westminster Gazette* (18 Dec. 1909), *Dial* (31 Jan. 1918); *London Mercury* (July 1922); Sassoon, *Best of Friends*, 61; Lillah McCarthy, 104; Hardy, *Literary Notebooks*, 119; Murry, 'Poetry', 132, 127; Murry, *Problem* 131.

Other reviews citing the 'transformation' of Hardy's verse are: *The Gownsman* (28 Apr. 1910) ('we soon perceive that we are in the hands of a master'); *Oxford and Cambridge Review* (Lent Term, 1910) ('our helpless and at times unwilling pleasure . . . is compelled by the presence of the indefinable thing which . . . turns lead to gold'); *Daily News* (26 Jan. 1917); *Westminster Gazette* (8 Dec. 1917) (Hardy 'works miracles'); *TLS* (13 Dec. 1917); *Pall Mall Gazette* (11 Nov. 1919); *Dorset County Chronicle* (2 June 1920) ('No poet, with the exception of Robert Browning, can so marvellously

In this book, I have tried to illuminate Hardy's metres by connecting them with Victorian metrical theory and with the development of sound symbolism in stanzaic verse. But the larger question of Hardy's ultimate success as a metrical poet will no doubt remain.

In the poem used as epigraph to this book, J. V. Cunningham hopes that chance will combine with craft to lend his poems grace. The great question in the study of Hardy's poetry is whether craft leads to grace and beauty and insight, or whether craft forestalls these things, leaving them to happen by accident in spite of craft. It may still be no defence of Hardy to say with Auden (above, Introduction) that Hardy's fondness for complicated stanza forms was 'an invaluable training in the craft of making'. Indeed Auden implies that this training was the adolescence of his own art, enabling him to go on to a maturer art. The literary tradition in general remains embarrassed by Hardy's commitment to craft and yet acknowledges the beauty and insight, and sometimes grace, of his poems.

Hardy was indeed committed to the craft of accentual-syllabic metre, and its assorted technicalities, in a most single-minded way. And there is a gap between his craft and his accomplishment; the accomplishment occurs, in

transmute ... the harshness of single words into captivating sweetness of connected diction'); *TLS* (1 June 1922) ('then the beauty of the whole takes you and flows back through the whole poem'). Several of the reviews complaining of Hardy's awkwardness also acknowledge this puzzling phenomenon: *Westminster Gazette* (11 Jan. 1899); *Daily Chronicle* (21 Dec. 1899); *Cosmopolitan* (Mar. 1899); *Academy* (14 Jan. 1899); *The Speaker* (24 Dec. 1898); *Daily News* (13 Dec. 1909); *Nation* (4 Feb. 1915); *Westminster Gazette* (8 Dec. 1917); *Land and Water* (13 Dec. 1917); *Saturday Review* (22 Dec. 1917); *TLS* (27 Nov. 1919); *Nation and Athenaeum* (13 Oct. 1928) (yet once the language is accepted, the poem seems 'inevitable'). Some remarkable modern testimonies to this sense of transformation in Hardy's verse are those by Cleanth Brooks and F. R. Leavis. Brooks: 'The poem that hung limp and awkward at one's earlier perusal suddenly comes to life, the apparent gaucheness reveals itself as sturdy grace and quiet strength.... Some limiting threshold is passed; a critical conversion point is reached' ('A Note' 68–9); again, Brooks: 'what is too often an inept dissonance finds resolution, from time to time, in harmonies much richer than the music of some of Hardy's contemporaries who never depart from a limiting decorum' ('Language', 404). Leavis: 'And again and again we have instances of the characteristic diction in which something has happened to the familiar medley of ingredients, so that we can no longer call it insensitive or gauche' ('Hardy the Poet', *Southern Review* (1940), 90).

Cunningham's words, '[i]f chance with craft combines / In the predestined space'. Sometimes the combination takes place, sometimes it does not in Hardy's verse. But Hardy makes us think again about the importance of craft, and a commitment to craft, in bringing about this combination. Week after week, year after year, he studied verse forms and experimented with them; he had a fascination for their technicalities, a fascination which made him take notes from his contemporary metrists. He played with the constituents of traditional verse forms, broke them down, and built them up again. He continually let his own thinking and imagining intertwine with these forms. He played with the expressive possibilities of accentual-syllabic metre, and constructed whole poems as metrical models of mental experience. As the years went by, he achieved more subtle ways of rendering this experience through the increasingly complex technicalities of his forms. He never seemed to question the bases of his art: 'Mechanic repetitions please' (' "Why do I?" ', iii. 157).

But he also leaves us with a renewed respect for the importance of a commitment to craft, almost as an allegory of the importance of a commitment to order despite the fact that there is a hiatus between order and the point of having order. I am not speaking so much of the existentialist rage for order. I am speaking of the faith that there is some point to following a scheme, to carrying out a train of thought, to finishing a project, to perfecting a method, even though it is unclear how these things contribute to the salvation of the world, or even to the enrichment of consciousness. If there is a craft for managing one's syllables, there is also perhaps a craft for managing one's thoughts. Hardy was not immune to wondering about these things, and indeed he wondered about them within the forms of his poems. *The Dynasts* might be seen as an allegory of craft and chance, craft turned into Destiny, which seeks some consummation of freedom:

> But—a stirring thrills the air
> Like to sounds of joyance there
>> That the rages
>> Of the ages

Shall be cancelled, and deliverance offered from the darts that were,
Consciousness the Will informing, till It fashion all things fair!

(p. 525)

Constitutionally, Hardy held this as a 'forlorn hope' (*Complete Poetical Works*, ii. 325). Yet his commitments imply that will, in Cunningham's terms, leads to art, and art leads to grace.

The question which the critical tradition has raised about Hardy is not unlike the question it has raised about accentual-syllabic metre. We have seen that the five-hundred-year discussion of metre kept sidetracking into inadequate notions of what the craft of accentual-syllabic metre entailed, notions that defined the craft too mechanistically or too liberally. The Victorians finally defined that craft in a way that was adequate to its mysteries. So persuasive was the Victorian insight that it became the (unacknowledged) basis of modern critical understanding of accentual-syllabic metre. But the Victorians could not explain how the craft produced excellent poetry. Nevertheless, their scholarship was led by the belief that the learning of the craft, and the formulation of its principles, would help lead to that poetry. Hardy is peculiarly their poet, their best student, their best example.

With the accentual-syllabic line, as with the Hardy poem, we enter an experience that is full of predictability, Cunningham's 'predestined space'; and we continually encounter the chance rhythms of speech and the chance possibilities of insight produced by this encounter. This is the chance that fascinated Hardy and held him to his craft. Hardy's experience in reading 'Lodge's poem to Rosaline' (above, Chapter 2) seemed to be a born-again experience in terms of his poetic vocation: he saw the point and the power of craft in itself, in that it means nothing in itself but without it very little of value can be attained. 'I had not thought what ghosts would walk / With shivering footsteps to my tune', Hardy wrote in 'On a Midsummer Eve' (ii. 177):

> I lipped rough rhymes of chance, not choice,
> I thought not what my words might be;
> There came into my verse a voice
> That turned a tenderer verse for me.

Accentual-syllabic metre then became Hardy's hobby and eventually his profession; he tinkered and laboured in its workshop, and produced minor and major work. He continually tried to perfect his craft, distinguishing its necessary weakness as craft from mistakes in craft. For Hardy, craft consisted not only in understanding the nature of his tools, as the Victorian prosodists understood them, but also in knowing the history of the use of those tools, and incorporating that history in his practice. There is limitation in Hardy's awareness, and in his craft; but this limitation is very close to the humility that makes grace possible.

Metrical Appendix

THIS Appendix attempts to note those instances where Hardy's stanza forms duplicate stanza forms used earlier in the literary tradition.

I have attempted to check all the sources we know Hardy to have used. These sources are more or less equivalent to a comprehensive anthology of the literary tradition; they include most of the major poets in the tradition, plus many anthologies and many minor poets. All of the examples in the Appendix are from poets which we know Hardy either read or referred to or knew. Good places to begin the search for Hardy's poetic influences are Pinion's *Hardy Companion*, 200–24 and Wright's *The Shaping of 'The Dynasts'*, chapter 2. I have also followed the leads suggested by the *Life*, the novels, notebooks, letters, lists of books Hardy owned, and books from his library.

In preparing this Appendix, I was helped by various other sources. Jakob Schipper's *Englische Metrik*, condensed and summarized in his *History of English Versification*, is a schematic outline of many common English metrical forms used up to the middle of the nineteenth century. After his thorough discussion of Old English forms, Schipper's account of 'Verse-forms common to the Middle and Modern English Periods' (*History*, 182; *Englische Metrik*, ii. 163) settles into a structural or synchronic account, and is thus not so useful as a diachronic or genealogical account. Also Schipper's sources, extensive as they are, do not include sources used by Hardy. Also important is the 'Metrical Index of Tunes' in *Hymns Ancient and Modern* (1906). In a 1910 article, L. T. Weeks gives statistics on the frequency of the various octaves and sestets, and their combinations, in about 6,000 English sonnets. He does not, however, give statistics for the various ways in which the rhymes of octave and sestet are interlocked, perhaps because instances of such interlocking are relatively few (140 of the total). Weeks's statistics include sonnets written after the 1860s; so it is difficult to tell, in cases of some rare forms, whether Hardy originated or followed the form. Most important are the metrical tables of Hardy's poems in Elizabeth Hickson's *The Versification of Thomas Hardy*. Though various of Hickson's readings are mistaken, her basic outlines are extremely valuable. Before using them, I revised them and incorporated the verse forms from *The Dynasts*, *The Queen of Cornwall*, and uncollected poems.

All that the Appendix presumes to do is list historically prior examples of stanza forms which Hardy might have known. That he knew his sources in all cases is unlikely. It is likely, however, that in some of the more complex parallels he was deliberately echoing the earlier examples. In some of these cases, Hardy's poems seem to be comments on the earlier poems. In many cases he may have been conscious of the general historical tradition of the stanza form, without being conscious of the particular examples I have listed. I have not excluded any parallel that I have been able to find, though in the case of very common forms I have listed only known early influences on Hardy. No attempt has been made to reduce the stanza forms to simpler elements, i.e. by reducing ballad stanzas to fourteeners, and so on. The full significance of the parallels must wait for someone else's study.

Because Hardy used more accentual-syllabic forms than any other poet, and because the range of his potential sources is so immense, this Appendix may serve as an extensive encyclopaedia, in alphabetical order, of metrical forms used by poets in the English metrical tradition. All major rhyming forms from the English tradition are listed—except Poulter's measure (but see $a^3b^3a^4b^3$ in the Appendix), limerick, and the Burns stanza ($a^4a^4a^4b^2a^4b^2$), though other examples from Burns are listed. Rime Royal and the *Rubáiyát* stanza are listed only because of a rhyme scheme parallel; and the Spenserian stanza is listed though Hardy's version of it no longer exists. Not included, because not used by Hardy, are certain romance forms (like ballade, *lai*, and sestina) and certain uncommon classical Greek and Latin forms imitated by poets in English. Hardy used blank verse in 'Domicilium' (iii. 279) and parts of 'Panthera' (i. 337) from the *Collected Poetical Works*, and also of course in *The Dynasts* and *The Queen of Cornwall*. In all cases I have tried to list the earliest predecessors I could find, plus an additional range of examples if such exist.

The Appendix of Hardy's forms constitutes an extraordinarily broad compendium of hymn forms, ballad forms, songs, and sonnet forms. Romance language imports are represented by *terza rima*, villanelle, rondeau, and triolet; Persian imports by ghazal and the Pantoum stanza; classical imports by sapphics, hendecasyllabics, alcaics, and some later accentual Latin hymn metres. In sum, approximately 170 forms, used by Hardy in about 500 poems, are listed here.

The general collections most often referred to in the following pages are:

1. *A New Version of the Psalms of David* by Tate and Brady, bound with Hardy's copy of the *Book of Common Prayer* (Cambridge, 1858), signed by Hardy and dated 1861. Like the readings in the Prayer Book, the psalms are copiously annotated by Hardy with names, places, and dates ranging from 1861 to 1868. Hardy claimed that his favourite hymnal was Tate and Brady's, much of which he knew by heart at age 10 (*Life*, 18). Describing the musical tradition of Hardy, his father, and his grandfather, the *Life* notes: 'In their psalmody they adhered strictly to Tate-and-Brady—upon whom, in truth, the modern hymn-book has been no great improvement' (*Life*, 10, see also 373–4, 393, 275). Tate and Brady's 'New Version' of 1696 was the successor to Sternhold and Hopkins's 'Old Version' which dates from Elizabethan times. On Hardy and Tate and Brady, see *Life*, 18, 275, 373, 393.

2. *Hymns Ancient and Modern*, 'the modern hymn-book'. The original edition was published in 1860–1 (with 273 hymns), and an appendix added in 1868 (for a total of 386 hymns). A revised edition was published in 1875 (with 473 hymns), with a supplement added in 1889 (an additional 165 hymns). This 1889 edition came to be called the 'Old Edition' and remained popular long after the revised edition of 1904 superseded it. The Dorset County Museum owns a copy of the 1889 edition, in which 221 hymns are marked, apparently in Hardy's hand; over 50 hymns are also listed by Hardy in the back of the book. Another copy of the *Hymns* owned by the DCM, presumably the edition of 1875 (the date of the preface), is also annotated but the hand does not look like Hardy's; pasted inside the front cover is a memorial card for Hardy's father, dated 31 July 1892. Though Hardy may have preferred Tate and Brady (only four of whose psalms are included in the 1889 *Hymns* (237–8, 249, 290)), he used *Hymns Ancient and Modern* often and imitated many of its forms. Colby College Library has *A Selection of Psalms and Hymns* (London, 1858), signed 'T. Hardy 1860–': most of the hymns are in standard metres, from Cowper, Tate, Brady, Watts, Wesley, and others. *Hymns Ancient and Modern* can be considered the climax of the hymnal tradition in the Anglican Church, for it marks the Church's official

acceptance of the place of hymns in church services. It also brings together the various traditions of hymnody, from versified psalms to personal poems. Finally, it contains a vast array of metrical forms, many of them borrowed from the secular tradition (which had, of course, often borrowed them from the sacred tradition). For an account of *Hymns Ancient and Modern*, see Frost.

3. Other hymn collections: John Keble's *The Christian Year*. In 1861 Hardy purchased an 1860 edition which is now in the Dorset County Museum. In this copy, which Hardy signed and dated 14 February 1861, many of the poems are marked or inscribed with dates and place names by him: i.e. Stinsford 1861, Fordington 1861, Kilburn 1862, London 1862, Clarence Place 1863, and so on. 'From September 1861, Hardy followed these weekly poems for Sundays and Feast Days' (Gittings, *Young Thomas Hardy*, 49). The dates in fact range from 1861 to 1866. The collection contains both traditional hymn forms and more original complex stanzas.

Hardy may also have been influenced by Sidney's 'Psalms', published in Hardy's copy of Grosart's 1877 edition of Sidney, and cited seven times as a possible source in the Appendix. Grosart includes Psalms 1–48. Sidney's psalms, along with those of his sister, the Countess of Pembroke, were important in linking the psalter tradition with the mainstream literary tradition. In their skilfulness and varied metrical forms, they take us out of the conventionality of the old psalters and into the tradition of Donne, Herbert, Crashaw, and Vaughan. This tradition in turn influenced the late development of a more original and personal hymnody as in Wesley, who adapted many of Herbert's poems, and in Keble (see above, Chapter 2).

4. Palgrave's *Golden Treasury*. Hardy admired this anthology more than any other (*Life*, 444). Urging Arthur Symons to put together an anthology of poems, he said: 'There has never been a good one since the first edition of the Golden Treasury—spoilt in subsequent edns' (*Collected Letters*, iii. 241). His own copy, now in the DCM, is a well-thumbed 1861 edition, signed 'T. Hardy from H. M. Moule Jan 1862'. Thirty-four of the poems are marked or inscribed by Hardy: 1–6, 8–11, 14, 31, 48 (3rd stanza dated 'Sept 24–73'), 49, 53, 60 ('critics' written next to 3rd stanza), 66, 82 ('My grandfrs song'), 107, 123, 140 ('mem: S.K. Museum—Etty. 1863' written at end of 3rd stanza), 146, 152 (dated 'Xmas '65'), 154–6, 158 (see below), 159, 164, 188, 217 ('R. Emmet' written in margin), 273, 285 ('i.e. the scene below' written in margin), 287; 16 poems are marked in the 'Index of First Lines': 31, 37, 59, 77, 80, 89, 90, 99, 106–9, 156, 160, 170, 194.

5. The music books used through the years by Hardy and his family. These include 'Books of Carols c. 1820', 'Book of Dance Music [*c.* 1820]', 'Books of Carols, Belonging to Thomas Hardy's Father [used to about 1842]', 'Music Book. 1799 Belonged to T.H.'s Grandfather', 'The Carol Book of T.H. I (1778–1837)', in the Dorset County Museum. To these might be added 'Words of Old Country Songs of 1820 Onwards. Compiled by Hardy about 1926', also in the DCM collection. These have been microfilmed on Reel 10 of *The Original Manuscripts and Papers of Thomas Hardy*. As a boy imitating his father on the violin, he 'was soon able to tweedle from notation some hundreds of jigs and country-dances that he found in his father's and grandfather's old books' (*Life*, 22).

The interrelationship between these secular and sacred sources constantly fuels Hardy's works. As the chronicler of a church choir which he had never seen 'as such, they ending their office when he was about a year old' (*Life*, 12), Hardy harked backed to his grandfather in *Return of the Native*: Thomasin's father, after playing secular music in a band, 'then, when they got to church door he'd throw down the clarinet, mount the gallery, snatch up the bass-viol, and rozum away as if he'd never played anything but a bass-viol' (I. 5, p. 53). Hardy noted in the 1896 preface to *Under the Greenwood Tree*: 'It was customary to inscribe a few jigs, reels, hornpipes, and ballads in the same book, by beginning it at the other end, the insertions being continued from front to back till sacred and secular met together in the middle, often with bizarre effect.' In this same preface, Hardy fondly remembered the old pedlar who would supply the church fiddlers with string and compositions: 'Some of these compositions which now lie before me, with their repetitions of lines, half-lines, and half-words, their fugues and their intermediate symphonies, are good singing still'.

6. *Ballad Minstrelsy of Scotland*, 2nd edition (Glasgow, n.d.). Hardy's copy, signed by him, is in the DCM. It has markings by him on pages 29, 31, 71, 74, 78, 80, 162, 167, 170, 224, 229, 240, 268, 270, 308, 321, 489, 492, 512. In the index, 'Bonnie Susie Clelland', 'Johnnie Armstrong', and 'Mary Colvine' are marked. The DCM also has Hardy's copy of Percy's *Reliques*, ed. Robert Willmott (London: *c.*1857) with 'King Estmere' marked in the text, and 'Admiral Hosier's Ghost' and 'Jemmy Dawson' marked in the index. These are only some of a number of sources of ballads which Hardy would have known.

7. *The Song Book: Words and Tunes from the Best Poets and Musicians*,

ed. John Hullah (London, 1866). Hardy's copy in the DCM has notes in his hand on pages 11, 23, 40, 56, 76–8, 80, 88, 94, 100, 102, 106, 114–15, 117–18, 127–8, 136–7, 140, 143, 279. After the rear index, Hardy writes: 'Dibdin pp. 62, 100, 104, 106': see below under $a^4a^4a^4b^3c^4c^4c^4b^3$. As late as 1907, Hardy recommended Hullah as a source of the 'old songs' (*Collected Letters*, iii. 247).

8. *English Lyrics: Chaucer to Poe 1340–1809*, ed. W. E. Henley (London, 1897). The DCM copy, signed 'Thomas Hardy', has markings on pages x, 5, 80, 136, 282, 314, 331.

9. *English Verse: Lyrics of the 19th Century*, ed. W. J. Linton and R. H. Stoddard (London, 1884). Hardy's copy, signed by him, is in the DCM. He marked 'Castles in the Air' by Thomas Peacock, 'Genius' and 'Dirge' by Richard Horne, and 'Alas!' by Phoebe Cary—all examples of Hardy seeking out minor poets in anthologies whose major entries he knew well.

10. *Ballades and Rondeaus*, ed. Gleeson White (London, 1887). Florence Henniker sent Hardy a copy in 1893 (*Collected Letters*, ii. 24). Hardy was impressed by the book and refers to it again in 1899 (*Collected Letters*, ii. 227). Hardy's perusal of it, in the years when he was resuming poetry, apparently led to his experimentation with the triolet, villanelle, Pantoum stanza, and rondeau.

11. *The Shakespeare Anthology 1592–1616 A.D.*, ed. Edward Arber (London, 1899). Hardy's copy is in the DCM. Hardy also owned Arber's *Dunbar Anthology 1401–1508 A.D.* (London, 1901).

12. *The Oxford Book of English Verse 1250–1900*, ed. Quiller-Couch (Oxford, 1900). At first reading, Hardy was 'much disappointed' in the volume (*Collected Letters*, ii. 277). In *Talks*, 7, Hardy says that he wishes Quiller-Couch would 'continue the *Oxford Book of English Verse* to 1900'. But the *Oxford Book*, first published in 1900, does include poems from the 1890s. The account in *Talks* seems to be garbled. Perhaps Hardy had said: 'continue the *Oxford Book of English Verse* 1250–1900', i.e. continue it into the twentieth century, something Quiller-Couch did not do until 1939 when he published the *Oxford Book of English Verse 1250–1918*.

13. *The Oxford Book of Victorian Verse*, ed. Quiller-Couch (Oxford, 1912). Hardy's copy in the DCM has a few markings.

Hardy's reading in the primary texts from the mainstream English poets was also extensive. He borrowed many of their metrical forms. He owned pre-1870 editions of Samuel Butler, Byron, Churchill, Coleridge, Cowley, Cowper, Crabbe, Dryden, Goldsmith, Herbert,

Lamb, Milton, Thomas Moore, Newman, Ossian, Otway, Shakespeare, Southey, Spenser, Swinburne's *Chastelard*, Thomson (Beattie *et al.*), Waller, Wordsworth, and Young—all of these in the Dorset County Museum; also Burns (Univ. of Texas), another Cowper (Holmes's list of Purdy books), Longfellow (Export Book Co. catalogue), Pope (Texas, Holmes's Listing), Scott (Holmes's Listing), Shelley's *Queen Mab and Other Poems* (Grolier catalogue). Except for the signed and dated copies, we do not know how many of these he purchased before 1870, but they suggest the kind of collecting he was beginning then. The Coleridge, Dryden (1865), Milton (1865), Thomson, and Shelley editions were signed by Hardy in the 1860s; during this decade Hardy refers to Wordsworth, *Childe Harold*, and *Lalla Rookh* (*Life*, 48–9, 58). Also in the early pages of the so-called '1867' notebook he made notations (possibly in the 1860s) from Swinburne's *Atalanta in Calydon* and Shakespeare, and later refers to his reading of Shakespeare, Spenser, Scott, Tennyson, Browning, Whitman, and Swinburne's *Poems and Ballads* during this period (*Life*, 47, 49, 345, 57–9). *The Golden Treasury* was, of course, an important resource for Hardy in the 1860s. *Desperate Remedies*, Hardy's first published novel, was begun in 1869 and alludes to several examples from Palgrave and also to Whitman, Coleridge, Moore, Scott, Tennyson, Browning, and Rossetti's *The Blessed Damozel*. Hardy's 'Studies, Specimens &c' notebook from the 1860s includes extracts from Barnes, Shakespeare, Byron, Wordsworth, Milton, Shelley, Swinburne, *The Golden Treasury*, Spenser, Burns, Tennyson, Jean Ingelow—on many of which Hardy makes technical observations (Hardy, *Literary Notebooks*, i. 267, 287, 345, 355, 361, 366, 379, 373, 375; Millgate, *Biography*, 87). In addition, Hardy recollected that he was 'much interested' in W. J. Mickle, 'when I was a boy' (*Collected Letters*, vi. 266).

In the 1860s Hardy was also reading books borrowed from friends like Moule and libraries like Mudie's. He spent time reading in the British Museum and in the Victoria and Albert Museum reading room. Hardy owned later editions of several of the above poets, and also (with the date of the editions given) Arnold (1890, 1893), Beaumont and Fletcher (1872), Barnes (1868—but given to Hardy in 1876; 1879, 1906), E. B. Browning (1887), Robert Browning (1893, 1897), Campbell (1904), Carew (1893), Chatterton (1875), Chaucer (1880), Clough (1871), Collins (*c.*1870), Davidson (1894), De La Mare (1906), Donne (1896), Drayton (1876, 1907), Fitzgerald's *Rubáiyát* (1909), Gray (1854—but signed 'Thomas Hardy. July 10,

1918'; 1885), Herrick (1893), Housman (1898), Jonson (1873), Keats (probably 1872—see Wright, 80), Kipling (1892, 1899), Masefield (1910, 1916), Meredith (1892), Henry Newbolt (1909), O'Shaughnessy (1874), Poe (1874–5), Schiller (1887), Sidney (1877), Surrey and Wyatt (*c*.1854–70), Tennyson (1875), Francis Thompson (1897), Vaughan (1897), Whitman (1912), Wycherley, Congreve, *et al.* (1871)—all of these in the Dorset County Museum; also Beddoes (*c*.1907, Texas), Blake (1901, Holmes list), Bridges (1905, Maggs catalogue), Emily Brontë (1923, Cox), Clare (1897, Cox), Darley (1908, Texas), Dowson (1905, Cox), Hood (1883, Texas), Rossetti's 'Hand and Soul' (Colby College), and Symons (1905, Cox). Not mentioned here are the many minor late Victorian and early twentieth-century poets Hardy knew and whose editions he owned. I have cited them below when they contain a metrical form which predates one of Hardy's.

Of course, Hardy was profoundly influenced by other aspects of earlier poems: for examples, see Marsden and Paulin. Also he was influenced by metrical forms which he imitated only in part. Thus, Thomas Hood's 'The Bridge of Sighs' (Palgrave, 231) probably influenced Hardy's 'Lonely Days' (ii. 429); and Swinburne's 'A Leave-Taking' probably influenced Hardy's 'A Broken Appointment'. But generally I have noted only those instances where the parallels are fairly precise, that is, where Hardy's stanza form corresponds to an earlier stanza form in (*a*) the number of lines in the stanza, (*b*) the rhyme scheme and the type of rhyme (masculine versus feminine), (*c*) the number of stress positions in each line, and (*d*) the distinctive rhythm (rising, falling, mixed). Thus I would not normally list Browning's 'Pisgah-Sights' in duple falling rhythm as a parallel to Hardy's 'Evelyn G. of Christminster' (ii. 393) in duple rising rhythm, though both follow the dimeter scheme: $a^2b^2a^2b^2c^2d^2c^2d^2$. (For a better parallel, see below.)

For approximately forty-six of Hardy's borrowed verse forms, we have unique prior examples. For the other Hardy forms listed here, we have a choice of two or more prior examples. The general index of this book may guide the reader to poets listed below. The texts most often cited as possible (but not necessarily exclusive) sources are Palgrave's *Golden Treasury* (33 times), *Hymns Ancient and Modern* (28), Linton and Stoddard (11), Heine in English translations (11), Henley's *English Lyrics* (11), Hullah's *Song Book* (9).

I think Hardy would have been fascinated by this Appendix and sympathetic to its aims. One of the effects of compiling it is the realization of the manifold ways in which metre shapes and is shaped by meaning, and the rich influence of the tradition of metrical forms on a given poem.

A particularly evident way in which Hardy's mingling of originality and tradition can be seen is in those poems which combine traditional stanza forms with original ones. We have seen how 'A Singer Asleep' makes the final *ottava rima* issue from the shreds and patches of earlier original stanza forms. A browse through the Appendix reveals many more instances of such metrical play. 'A Bird-scene at a Rural Dwelling' (iii. 7) moves from a complex stanza into a pentameter triplet. 'In a Waiting-Room' (ii. 266) moves similarly into a common hymnal form. Some poems, like 'Shut out that Moon' (i. 265), 'Misconception' (i. 283), 'The Haunter' (ii. 55), and 'The Rover Come Home' (iii. 127) enact the emergence of a more regular hymnal form out of a rougher ballad or original form—like a rover come home. 'Drawing Details in an Old Church' (ii. 475) and 'Bereft, She Thinks She Dreams' (ii. 97) reverse this movement, and awaken us from the conventional form. Similarly, 'In Sherborne Abbey' (iii. 74) begins with an heroic quatrain and moves into more complex stanzas. 'I met a man' (ii. 304) brackets its series of complex stanzas with tetrameter triplets. 'Every Artemisia' (iii. 31) and 'The Flower's Tragedy' (iii. 103) alternate complex stanza forms with tetrameter couplets. 'Drinking Song' (iii. 247) does this also but concludes with a tetrameter triplet. 'The Going' (ii. 47) alternates a Chatterton-like stanza with original stanzas. 'The Chapel-Organist' (ii. 406) begins and ends with a stanza in alexandrine couplets, and the remaining stanzas are a kind of alexandrine blank verse each ending with a couplet. Several of the poems in 'Satires of Circumstance' (ii. 140–9) seems almost to satirize conventional forms; some are in upside-down sonnet forms. One indeed ('Over the Coffin' ii. 148) gives us a common English tetrameter sestet in the first stanza, then mimics an *ottava rima* (in tetrameters!) in the second stanza, and the whole adds up to a tetrameter sonnet with the sestet coming first. And within this hall of metrical mirrors, the characters play out their drama 'over the coffin'. How such play with original and traditional form intersects with meaning is of enormous interest in Hardy, and can be seen at work throughout the Appendix.

Couplet Stanzas

a^2a^2
d.r.

An early English form, adapted in Skelton's dimeter couplets; represented in Drayton's 'An Amouret Anacreontick' and Herrick's 'His Almes'.

Alternate stanzas in Hardy's 'Signs and Tokens', ii. 275; final stanza of 'The Choirmaster's Burial', ii. 284; used in continuous long stanza in '"Quid Hic Agis?"', ii. 175; various long stanzas in 'The Clock-Winder' (d.t.r.) ii. 268; used in the balance of 'Thoughts at Midnight' (d.f.), iii. 168.

a^4a^4
d.r.

Octosyllabic couplets, the oldest of the English forms now in use (Saintsbury, i. 57). Separated into couplet stanzas in ballad measures like 'Willie's Lady' (Child, *English and Scottish Popular Ballads*, 6A). Other examples Hardy knew: Herbert's 'Charms and Knots'; Browning's 'The Boy and the Angel' in *Selections from the Poetical Works* (London, 1893), in Hardy's DCM collection with some forty poems marked by him; Heine's 'Belshazzar' trans. Bowring (see below under $a^3b^3a^3b^3$ d.r.) and trans. Leland. Also see below, $a^4a^4b^4b^4$, $a^4a^4b^4b^4c^4c^4$, and $a^4a^4b^4b^4c^4c^4d^4d^4$.

Hardy's 'The Abbey Mason', ii. 124, 'The Paphian Ball', ii. 137, written in what Hardy called a 'familiar mediaeval mould' (*Complete Poetical Works*, iii. 319) and *Dynasts*, 53, which varies the rhythm; alternate stanzas in 'Memory and I', i. 226, 'Every Artemisia', iii. 31, 'The Flower's Tragedy', iii. 103; occasional stanzas in 'The Three Tall Men', iii. 187 and 'Drinking Song', iii. 247; first stanza in 'A Wife Comes' Back', ii. 368; final stanza in 'The Love-letters', iii. 173 and 'A Nightmare, and the Next Thing', iii. 202;

used in stanzas of various length in 'Lady Vi', iii. 119; extended into a 12-line stanza in 'The Sweet Hussy', ii. 109, and a 20-line stanza (d.f.) in 'The Felled Elm and She', iii. 204. In trisyllabic rhythm, see 'Cardinal Bembo's Epitaph on Raphael', i. 224.

a^5a^5
d.r.

Heroic Couplets, common early English form; used also in *Hymns* (1889) 'Peace, Perfect Peace', 537, marked in Hardy's edition. (See also $a^5a^5b^5b^5$ below.)

Hardy's 'The Coronation', ii. 88; first and last stanzas of 'An Unkindly May', iii. 174. 'Panthera', i. 337 uses such couplets, undivided into stanzas, in the first two and last stanzas. There are several individual couplets in *The Dynasts* and the *Queen of Cornwall*. The form is extended into a 12-line stanza in 'A Christmas Ghost-Story'. i. 121. Used in stanzas of various length in 'Lines', i. 104.

a^6a^6
d.r.

Alexandrine couplet, early English form, to be used in Drayton's *Polyolbion*; to be distinguished from classical hexameter imitations.

Hardy's 'An Evening in Galilee', iii. 216 in long stanzas; similarly in 'The Chapel-Organist', ii. 406, in first and last stanzas, and in concluding couplets of other stanzas.

a^7a^7
d.r.

The venerable septenary or fourteener couplet, a favourite of Elizabethan translators like Chapman, Golding, and others. Other examples are Byron's '"There's not a joy"' (Palgrave, 222), Richard Edwardes's 'Amantium Irae' in *Oxford Book of English Verse*.

Hardy's 'The Homecoming', i. 303 (in couplets and quatrains)

Triplets

a³a³a³
d.r.

Skelton's 'In Praise of Isabel Pennell', in Henley.
Hardy's 'To Sincerity', i. 336.

a⁴a⁴a⁴
d.r.

Keble's 'Trinity Sunday' marked 'Kilburn 62, W.P.V. 66' by Hardy; Palgrave, 93 (Herrick's 'Whenas in Silks'); Herrick's 'You are a tulip'.
Hardy's '1967', i. 269, 'Tolerance', ii. 42, 'A Week', ii. 95, 'To C.F.H.', iii. 134; *Dynasts*, 118, 301, 306, 322, plus others of more rhythmic variety: 'In the Moonlight', ii. 148, 'The Dead and the Living One', ii. 300, 'Starlings on the Roof', ii. 108, and 'Inscriptions for a Peal of Eight Bells', iii. 122. Extended into a 12-line stanza, $a⁴a⁴a⁴b⁴b⁴b⁴$ etc., in 'Horses Aboard', i. 105. First and last stanzas of ' "I met a man" ', ii. 304; last stanza of 'Whispered at the Church-opening', iii. 241 and 'Drinking Song', iii. 247; 3rd stanza of 'A Nightmare, and the Next Thing', iii. 202.

a⁵a⁵a⁵
d.r.

Gosse's 'Lying in the Grass' from *On Viol and Flute*, which Hardy owned in the 1890 edition. Hardy referred to this volume and to *Firdausi in Exile* (see below, $a⁴b²a⁴b²$) as 'your two compact and teeming volumes' (*Collected Letters*, ii. 208); John Hay's 'A Woman's Love', in Linton; Rochester's 'To His Mistress', in Henley.
Hardy's 'At the Royal Academy', ii. 400, 'Childhood Among the Ferns', iii. 199, *Dynasts*, 180, 198, 215, 312, 353, *Queen of Cornwall*, XXIII, p. 72, IX, p. 33; 'A Bird-scene at a Rural Dwelling', iii. 7 (last stanza).

a⁴b⁴a⁴
b⁴c⁴b⁴
etc.
d.r.

Terza rima used in Browning's 'The Statue and the Bust' concluding with a quatrain. Hardy thought it 'one of the finest of Browning's poems' (*Life*, 199).

Hardy's 'In a Cathedral City', i. 272; ' "She charged me" ', ii. 78, 'The Jubilee of a Magazine', ii. 134, 'The Dead Bastard', iii. 209, all concluding with a quatrain; 'George Meredith', i. 358 (ending in a triplet, like *Dynasts*, 483), 'A Plaint to Man', ii. 33 (ending in a couplet, like *Dynasts*, 298).

$a^4b^4a^4$
$a^4b^4a^4$
etc. with
concluding
quatrain $a^4b^4a^4a^4$
19 lines in all
d.r.

Villanelle as in several examples in White: Dobson's 'For a Copy of Theocritus', 247 and others, Gosse's 'Wouldst thou not be content to die'. White said: 'The VILLANELLE has been called "the most ravishing jewel worn by the Muse Erato" ' (lxxiii). On White, see above, p. 56.
Hardy's 'The Caged Thrush Freed and Home Again', i. 184.

$a^5b^5a^5$
$b^5c^5b^5$
etc.
d.r.

Terza rima in traditional Dantesque form (Schipper, ii. 895–7). Shelley's 'Ode to the West Wind' (Palgrave 275), which ends with a couplet, is a prominent example.
Hardy's 'The Burghers', i. 31, with concluding triplet like the following; Hardy's *terze rime* in other forms include 'To a Bridegroom', iii. 281 $a^4b^4a^3$; 'Friends Beyond', i. 78, $a^8b^4a^8$ d.f.; 'The Letter's Triumph', iii. 239, $a^4b^5a^3$. Many other Hardy poems use interlocking rhymes schemes between stanzas. On the general subject of such rhyme-linking, see Hickson, 45–50.

Quatrains

$a^4a^4a^4b^2$
d.r.

Palgrave, 21 (Wyatt's 'Forget Not Yet') with the *b* line repeated as a refrain.
Hardy's 'Dream of the City Shopwoman', ii. 379 (with the same *b* rhyme every 2 stanzas).

$a^4a^4b^2a^4$
d.r.

Symons's 'For a Picture of Rossetti', from *A Book of Twenty Songs* (London, 1905), which Hardy owned. Unlike Hardy's lines, all Symons's lines are headless.
Hardy's ' "He inadvertently cures his love-pains" ', iii. 118; others, 'A Conversation at Dawn', ii. 80, 'A Military Appointment', ii. 455, in more trisyllabic rhythm. This is the *Rubáiyát* rhyme scheme.

$a^4a^4b^4b^4$
d.r.

A common form in *Hymns* (1861), many of which are marked by Hardy in his 1889 *Hymns*; 'Awake, my soul' (3), one of Hardy's three favourite hymns (*Life*, 275) and the first hymn in *Hymns* (1861), is in this form; also Tate and Brady 57, 60; Palgrave, 5 (Marlowe's 'The Passionate Shepherd to His Love'), partly underlined in Hardy's copy; Palgrave, 95 (Waller 'On a Girdle') and others; 'Cospatrick', marked by Hardy in *Ballad Minstrelsy*; Heine's 'She Dances' on p. 396 marked by Hardy in Bowring's trans. (See above, $a^3b^3a^3b^3$.) Also see Schipper, ii. 469–70.
Hardy's 'The Two Men', i. 100, 'A Young Man's Epigraph on Existence', i. 359, 'A Poet', ii. 138, 'The Garden Seat', ii. 331, 'The Fading Rose', iii. 89; *Queen of Cornwall*, I, p. 7; VII, p. 28; *Dynasts*, 522; other poems with more rhythmic variation are 'The Newcomer's Wife', ii. 79 and 'Christmas: 1924', iii. 256; last stanza of 'On the Death-Bed', ii. 147 and 'In a Waiting-Room', ii. 266. Also see $a^4a^4b^4b^4c^4c^4d^4d^4$.

$a^4a^4b^4b^4$
t.r.

Common Song measure: for example 'While Shepherds Were Feeding' in Hardy's 1820 Book of Carols in *Original Manuscripts*; Palgrave, 251 (Wordsworth's 'The Reverie of Poor Susan'), 143 (Cowper's 'The Poplar Field'), 169

(Byron's 'O Talk Not to Me'), and others;
Shelley's 'The Sensitive Plant', Browning's
'The Laboratory', Byron's 'Song of Saul' in
Linton, Moore's 'Oh! Breathe Not His
Name', marked by Hardy in Hullah, 279;
'The Lost Lady', a Dorset song cited by
John Udal, 325. Udal also gives other
examples of Dorset songs Hardy would
have known.

Hardy's 'The Ruined Maid', i. 197, 'The
Orphaned Old Maid', i. 296; 'Suspense',
iii. 232; *Dynasts*, 312.

$a^4a^4b^4b^4$
d.f.

Shakespeare's songs in *Tempest*, IV. i. 106
and *Love's Labour's Lost*, IV. iii. 101 (not
divided into stanzas); Thomas Campbell's
'How Delicious is the Winning' (Palgrave,
183): *Under the Greenwood Tree's* chapter
title v. 1, quotes a line from this poem. Also
see Palgrave, 288 (Shelley's 'Music, When
Soft Voices Die').

Hardy's 'The Reminder', i. 324.

$a^5a^5b^5b^3$
d.r.

Swinburne's 'A Wasted Vigil' from *Poems
and Ballads*, Second Series, which Hardy
owned in an 1887 edition (DCM). Unlike
Hardy's, Swinburne's fourth line is
repeated as a refrain.

Hardy's 'The Old Workman', ii. 441.

$a^5a^5b^5b^5$
d.r.

Heroic couplets in quatrains, as in *Hymns*
(1889), 252 and 312, marked by Hardy;
'To the Goddess of Love', Wharton's
translation in his edition of *Sappho*, in
Hardy's DCM collection: also see below,
$a^3b^3a^4b^2$.

Hardy's 'The Graveyard of Dead Creeds',
iii. 33 (last stanza); 'In Sherborne Abbey',
iii. 74 (first stanza); Hardy's *Dynasts*, 153,
310, 398; 'Rome: The Vatican: Sala delle
Muse', i. 136 with much trisyllabic
rhythm, and some stanzas of 'Jubilate', ii.
257. Also, 'Motto for the Wessex Society of
Manchester', iii. 294.

$a^7a^7b^7b^7$
d.t.r.

Ainsworth liked this ancient form, for example in 'The Legend of the Lady of Rookwood' and others in *Rookwood*; Campion's 'Never Weather-Beaten Sail' in Henley (136) (though Hardy's reading it here postdates 'In Tenebris II'); Byron's 'There's Not a Joy' (Palgrave, 222), Browning's 'Martin Relph' and 'The Bean-Feast'.
Hardy's 'In Tenebris II', i. 207, 'Wessex Heights', ii. 25, 'The Market-Girl', i. 292, 'The Elopement', ii. 93, 'The Mock Wife', iii. 77. Also see 'Our Old Friend Dualism', iii. 233 ($a^7a^7b^7b^7c^7c^7$).

$a^2b^2a^2b^2$
d.r.

Thomas Moore's 'By the Fair and Brave' in *Lalla Rookh*; Henry Newbolt's 'A Sower' (d.r.).
Hardy's 'The Wound', ii. 202; 'The Pine Planters', I (i. 328) (stanzas 6, 7, 8).

$a^3b^2a^3b^2$
d.r.

Heine's 'The Roaring Waves are Dashing' on p. 107 marked by Hardy in Bowring. (See below, $a^3b^3a^3b^3$.) Moulton's 'The Ghost's Return' in *In The Garden of Dreams* (see $a^4b^3c^4b^2$ below); Locker's 'Heine to His Mistress' in *London Lyrics* (on Locker, see $a^4a^4b^2c^4c^4b^2$ d.r.).
Hardy's 'The Weary Walker', iii. 54. Also see $a^2b^2a^2b^2c^2d^2c^2d^2$. The quatrain is joined with a hymn quatrain in 'Drawing Details in an Old Church', ii. 475, and extended to a sestet ($a^3b^2a^3b^2a^3b^2$) in 'The Wanderer', ii. 367. Other poems in this scheme in trisyllabic rhythm are: 'In Time of "the Breaking of Nations"', ii. 295. 'A Gentleman's Epitaph', ii. 350. 'Without, Not Within Her', ii. 423, 'The Church and the Wedding', iii. 99, and 'Cynic's Epitaph', iii. 115.

$a^3b^3a^3b^2$
d.r.

Richard Stoddard's 'Under the Rose' in Linton, 259.

Hardy's 'A January Night', ii. 204; 'The Selfsame Song', ii. 367; ' "We say we shall not meet" ', iii. 223; and every third stanza of 'Haunting Fingers', ii. 357. In trisyllabic rhythm: 'Her Temple', ii. 401.

$a^3b^3a^3b^3$
d.r.

Hymns (1889), 224, 'O Happy Band', marked by Hardy, and 217, 'Thy Kingdom Come', both in the earlier editions; a form frequently used in E. A. Bowring's translation of Heine's *Poems* (London, 1878), which Hardy owned (Purdy, 117), i.e. pp. 106 ('The Maid Stood by the Ocean') and 217 ('I Fain Would Linger by Thee') both marked by Hardy. For Heine also see a^4a^4, $a^4a^4b^4b^4$, $a^3b^2a^3b^2$, $a^4b^3a^4b^3$, $a^4b^4a^4b^4$, $a^4b^4b^4a^4$, $a^3b^3c^3b^3$, $a^4b^4b^4a^4C^2$, $a^4a^4b^3c^4c^4b^3$, $a^4a^4b^4c^4c^4b^4$; 'All in a Garden Green' in Hullah, 124; also 'The Lover Refused', and Herrick's 'To Live Merrily' in Henley (24, 179), though Hardy's reading them here postdated 'In a Eweleaze near Weatherbury'.

Hardy's 'Song from Heine', i. 223; also 'After the Club-Dance', i. 292, 'The Husband's View', i. 301, 'At Day-Close in November', ii. 43, ' "I looked up from my writing" ', ii. 305, 'The Little Old Table', ii. 425, 'The Last Time', ii. 472, 'The Protean Maiden', iii. 125, 'That Moment', iii. 141, all with some rhythmic variation. Also see $a^3b^3a^3b^3c^3d^3c^3d^3$. For eight other examples with more rhythmic variety, see Hickson, 94. These merge by degrees with the following.

$a^3b^3a^3b^3$
t.r.

Swinburne's 'An Interlude', marked by Hardy in his 1873 edition of *Poems and Ballads* i.
Hardy's 'The Fiddler', i. 200, 'The Old Neighbour and the New', ii. 458.

a³b³a⁴b²
d.r.

J. H. Merivale's translation of Sappho's 'The silver moon is set' in Wharton's *Sappho*, in Hardy's collection.

Hardy's 'While Drawing in a Churchyard', ii. 287, 'Julie-Jane', i. 298, though both are in increasingly trisyllabic rhythm.

a³b³a⁴b³
d.r.

'Short' Hymnal Stanza, or 'resolved poulter's measure' (Saintsbury, i. 328). (Poulter's measure is an English Renaissance form which alternates alexandrines and fourteeners: a⁶a⁷.) 4 hymns in this 'short' metre are marked in Hardy's 1889 *Hymns*: 120, 205, 270, and 284 'Far From My Heavenly Home' (this last initialed 'H.M.M.').

Hardy's ' "I look into my glass" ', i. 106. Others, 'To Life', i. 152, 'The Man He Killed', i. 344, 'The Wedding Morning', ii. 374, 'Intra Sepulchrum', ii. 469, vary the strict iambic measure.

a³b⁵a³b⁵
d.r.

William Jones's 'What Constitutes a State', Guest, 664; Emily Brontë's 'No Coward Soul is Mine' in the *Oxford Book of Victorian Verse*—but Hardy's reading it here postdates his writing of 'The Milkmaid'.
Hardy's 'The Milkmaid', i. 195.

a⁴b²a⁴b²
d.r.

Swinburne's 'In Memory of Walter Savage Landor', which Hardy marked in his 1873 edition of *Poems and Ballads* I; Ben Jonson's 'Epitaph on Salathiel Pavy' (not divided into stanzas), in the *Oxford Book of English Verse*, though Hardy's reading it here may postdate his writing of 'Misconception'.

Hardy's 'A King's Soliloquy', ii. 87, 'The Occultation', ii. 200; also see *a⁴b²a⁴b²c⁴d²c⁴d²*; one such quatrain is combined with a hymn stanza in 'Misconception', i. 283 and 'Bereft, She Thinks She Dreams', ii. 97.

$a^4b^3a^4b^2$
d.r.

De La Mare's 'Unregarding'; *Hymns*, 207, 'Our Blest Redeemer', marked by Hardy. Hardy's 'Not Known', iii. 260.

$a^4b^3a^4b^3$
d.r.

'Common Metre'. See below, ballad metre, $a^4b^3c^4b^3$. One of the earliest forms Hardy knew in Tate and Brady, and the 1861 *Hymns Ancient and Modern*. Twenty-one hymns in this metre are marked in Hardy's 1889 *Hymns*, including 165 marked 'Ps XC. paraphrased by Dr. Watts' and discussed in *Life*, 393; Palgrave, 82 (Herrick's 'Gather Ye Rose-buds'), next to the title of which Hardy wrote 'My Grandf'rs song'; also Palgrave, 83 (Lovelace's 'To Lucasta') and others.

Hardy's 'The Comet at Yell'ham', i. 188, 'Geographical Knowledge', i. 345, ' "It never looks like summer" ', ii. 253, and 'Epitaph on a Pessimist', iii. 124—all observe a strict iambic norm. Thirty-five other poems observe the form but with rhythmic variation as in ballads: see Hickson, 95, 97 and also 'The Sound of Her', first published in Hynes's edn., iii. 304. Also see below, $a^4b^3a^4b^3c^4d^3c^4d^3$.

$a^4b^3a^4b^3$
t.r.

Moore's 'Believe Me' from *Irish Melodies*; Heine's 'I Dreamt' on a page (206) marked by Hardy in Bowring (see above, $a^3b^3a^3b^3$). Ingelow's 'You Bells in the Steeple' from 'Songs of Seven'; Darley's 'Here's a Bank' from *Sylvia, or, the May Queen*, 139 (see also 432).
Hardy's 'Rose-Ann', ii. 302, 'Growth in May', ii. 398.

$a^4b^3a^4b^3$
d.f. (with Latin tag lines used in *b* lines)

First 4 lines of the Egerton manuscript, 'Of on that is so fayr and bright' (quoted in Saintsbury, i. 86). I don't know if Hardy knew this poem in the 1870s when he wrote the following.

Hardy's 'After Reading Psalms xxxix, xl, etc.,', ii. 484. Hardy uses the Latin tag only in the fourth line.

abab
bcbc etc.
with concluding
quatrain nana
d.t.r.

Pantoum stanza. 5 Pantoum poems, in either tetrameter or trimeter lines, are quoted in Gleeson White's *Ballades and Rondeaus*. White says of some of his examples: 'In Mr. Austin Dobson's "In Town" and Mr. Brander Matthews' "En route"—as the latter himself points out in *The Rhymester*—"there is an attempt to make the constant repetitions not merely tolerable but subservient to the general effect of monotonously recurrent sound— in the one case the buzzing of the fly, and in the other the rattle and strain of the cars' (lvii).

Hardy's 'The Second Night', ii. 439 uses a Pantoum rhyme scheme, in alternating tetrameter and trimeter lines, but does not repeat lines as in the orthodox fashion.

$a^4b^4a^4b^2$
d.r.

Hymns (1868), 269, 'St Gabriel'; Newman's 'O Thou, of Shepherds' (see below, $a^2b^2a^2a^2b^2$); Palgrave, 118 (Pope's 'Happy the Man'); Byron's 'On This Day I Complete My Thirty-sixth year'; Ingelow's 'The Letter L'; Herbert's 'Vertue'.

Hardy's 'On the Department Platform', i. 271; 'Life Laughs Onward', ii. 201; others with more rhythmic variation are: 'The House of Hospitalities', i. 255, ' "Nothing matters much" ', iii. 143, 'The Gap in the White', iii. 262.

$a^4b^4a^4b^3$
d.r.

Hymns (1889), 119, 'His are the Thousand Sparkling Rills' initialled 'E.L.H.' and listed in the back by Hardy; Ingelow's 'Song of the Going Away'.

Hardy's 'The Dead Quire', i. 310; others, 'A Circular', ii. 58, ' "I travel as a phantom now" ', ii. 194, with more rhythmic variation. Hardy's earliest use of the form, 'Her Father', i. 273, predates the 1889 hymnal.

a⁴b⁴a⁴b⁴
d.r.

'Long' Hymn Stanza. A common form in the early hymnals. Fourteen hymns in this form are marked in Hardy's 1889 *Hymns*. Palgrave, 72 (Wotton's 'Character of a Happy Life'), 86 (Darley's 'It is Not Beauty I Demand'), and others.
Hardy's 'The Rambler', i. 325 observes the strict iambic norm. Thirteen others vary this rhythm as in ballads: see Hickson, 96. Also second stanza of 'Under the Waterfall', ii. 45; third stanza of 'Whispered at the Church-opening', iii. 241.

a⁴b⁴a⁴b⁴
t.r.

Hymns, 637, 'Oh! Come to the Merciful Saviour', marked by Hardy; 'When the Rosebud of Summer', in Hullah, 140, marked by Hardy; Campbell's 'The Soldier's Dream' (Palgrave, 267).
Hardy's 'The Lizard', iii. 304. Two other poems, 'The Roman Gravemounds', ii. 116, 'A Question of Marriage', iii. 238, mix an iambic pace with this song lilt.

a⁴b⁴a⁴b⁴
d.f.

Several songs in Hullah: 'Care, Thou Canker', 'Shepherds I have Lost My Love', 'Phillis, Talk No More of Passion', and 'Cease, Rude Boreas' (these last three marked by Hardy, and the last inscribed 'Jenny Phillips' by him). See Hullah, 103, 143, 137, 118. Heine's 'Fast is Creeping on Us Dreary' on a page (122) marked by Hardy in Bowring (see above, $a^3b^3a^3b^3$).
Hardy's 'The Bridge of Lodi', i. 139.

a⁴b⁴a⁴b⁴
t.f.

Dryden's 'After the Pangs of a Desperate Lover' from *An Evening's Love* (see Saintsbury's praise of this lyric, ii. 347); Darley's 'Lullaby' from *Sylvia, or, the May Queen*, 188.

Hardy's 'The Voice', ii. 56. Peter Coxon notes an interesting echo in Hardy's first line to Horace's 'Eheu fugaces, Postume, Postume' (Ode, ii. 14); but the poem's metrical likeness to the alcaic stanza form is problematic: see below, on alcaics.

$a^5b^2a^5b^2$
d.r.

Swinburne's 'Lines on the Monument of Giuseppe Mazzini' from *A Midsummer Holiday and Other Poems*, owned by Hardy in the third edition (London, 1904—DCM) (Hardy discusses Swinburne and Mazzini in *Life*, 37); Lord Houghton's 'Shadows', part v ('Twould seem the world were large enough') marked by Hardy in his copy of Houghton (see *Collected Letters*, ii. 34; also see $a^4b^2a^4b^2c^4d^2c^4d^2$, $a^3b^3c^3b^3d^3e^3f^3e^3$); also the first and last stanzas of 'Come, Shepherd Swains' in *The Shakespeare Anthology*.
Hardy's 'Former Beauties', i. 291.

$a^5b^3a^5b^3$
d.r.

Vaughan's 'Corruption' (though not divided into stanzas), marked by Hardy (and given the marginal comment, 'Cf Wordsworth's Ode on Immortality') in his copy of Vaughan (Wright, 15).
Hardy's 'A Meeting with Despair', i. 75, 'The Peasant's Confession', i. 40, 'The Caged Goldfinch', ii. 234, 'A Beauty's Soliloquy during Her Honeymoon', iii. 115; others, 'One We Knew', i. 331, 'Near Lanivet, 1872', ii. 168, 'A Daughter Returns', iii. 245, and 'Lying Awake', iii. 198 with more varied rhythms.

$a^5b^5a^5b^2$
d.r.

Grant Allen's 'Forget-Me-Not' from *The Lower Slopes* (London, 1894—Hardy's copy, Univ. of Texas) (see below, $a^4b^4a^4a^4b^2$); 'My first borne love' from *The Phoenix Nest*; Sidney's 'Psalm 6'.
Hardy's 'After the Last Breath', i. 326,

'The Inscription', ii. 460 with more trisyllabic rhythms.

a⁵b⁵a⁵b³
d.r.

Ingelow's 'Honors'. F. H. Doyle's 'The Loss of the "Birkenhead"' in the *Golden Treasury*, Second Series, 93: 'I have been looking also into the second Golden Treasury of Palgrave' (*Collected Letters*, ii. 208); Tennyson's 'A Dream of Fair Women', Arnold's 'Progress'.
Hardy's 'The Colonel's Soliloquy', i. 117, 'The Ballad-Singer', i. 291, 'Henley Regatta', iii. 215; *Dynasts*, 512.

a⁵b⁵a⁵b⁵
d.r.

Elegiac Stanza or Heroic Quatrain as in Gray's 'Elegy' (Palgrave, 147); also Palgrave, 276 (Wordsworth's 'Suggested by a Picture of Peele Castle'); Arnold's 'Self-Deception'.
Hardy's 'Her Dilemma', i. 16, 'Spectres That Grieve', ii. 37, 'The Young-Glass Stainer', ii. 282, 'At Lulworth Cove a Century Back', ii. 371, 'Standing by the Mantlepiece', iii. 226; third stanza of 'The Graveyard of Dead Creeds', iii. 33; also 'In a Museum', ii. 163 in mainly triple rising rhythm. *Dynasts*, 34, 295, 348, 383, 393. Other examples in more varied rhythms: 'Aquae Sulis', ii. 90, 'Jezreel', ii. 333, 'A Wet August', ii. 343.

a³b⁵b⁵a³
d.r.

Felicia Hemans's 'The Song of Night'. Some contemporary citations of this well-known poet of the century are made in *Pelham*, an early favourite of Hardy (*Life*, 52) (Lord Vincent admires Hemans's poems, ch. 67) and *Wives and Daughters*, chs. 6 and 23. Three of Hemans's poems are included in Ward.
Hardy's 'A Young Man's Exhortation', ii. 370.

$a^4b^4b^4a^3$
d.r.

E. B. Browning's 'The Deserted Garden':
On Hardy and E.B.B., see *Life*, 192 and
Gittings, *Young Thomas Hardy*, 85.
Hardy's 'Neutral Tones', i. 13, and last
stanza of 'A Countenance', iii. 200.

$a^4b^4b^4a^4$
d.f.

Shakespeare's 'The Phoenix and the
Turtle'; Heine's 'When I'm With Thee,
Strife and Need' on p. 122 marked by
Hardy in Bowring (see above, $a^3b^3a^3b^3$
d.r.).
Hardy's 'He Wonders about Himself', ii.
256, with trisyllabic substitutions and
hypercatalectic lines.

$a^4b^4b^4a^4$
d.r.

Tennyson's *In Memoriam* stanza; for others,
see Schipper, ii. 546, and Saintsbury, ii.
158.
Hardy's 'The Torn Letter', ii. 19; second
stanza of ' "What did it mean?" ', ii. 431.

$a^5b^5b^5a^5$
d.r.

Italian quatrain.
Hardy's 'The Graveyard of Dead Creeds'
(first two stanzas), iii. 33.

$a^2b^2c^2b^2$
d.t.r.

Child's *Popular Ballads*, 210 'Bonnie James
Campbell'.
Hardy's 'The Pine Planters' I (first five
stanzas), i. 328, 'Boys Then and Now', iii.
226; *Dynasts*, 429–30.

$a^3b^3c^3b^3$
d.r.

Hymns (1861), 142, 'Brief Life is Here Our
Portion', marked in Hardy's 1889 edition
(225); also in *Hymns* (1889), 115 and 324
(d.f.) are marked. Also 'There Lies the
Glow of Summer' in Bowring's Heine, on
p. 82 marked by Hardy, with other
examples on pp. (marked by Hardy) 107,
186, 208, 213. This metre is also used often
(though uncertainly) by Leland in his
translation of Heine, a translation much
annotated by Hardy (see above, $a^3b^3a^3b^3$
d.r.).

Hardy's 'The Brother', iii. 218. In trisyllabic rhythms: 'The Death of Regret', ii. 114; extended to a sestet, $a^3b^3c^3b^3c^3b^3$, in 'Unkept Good Fridays', iii. 175; doubled into $a^3b^3c^3b^3d^3b^3e^3b^3$ in the first two stanzas of 'In the Small Hours', ii. 424.

$a^4b^3c^4b^2$
d.r.

Hymns, 254 (marked 'To ELH') and 499 (marked with a place name and dated); Louise Chandler Moulton's 'Do Not Grieve' in *In the Garden of Dreams* (London, 1890—Hardy's copy, Holmes Listing), which Hardy acknowledged in 1890 as her 'beautiful "Garden of Dreams" . . . I have read a good many of the poems—nearly all of which are penetrated by the supreme quality, emotion' (*Collected Letters*, i. 208) (see also $a^3b^2a^3b^2$ d.r. and $a^4b^2a^4b^2a^4b^2$); De La Mare's 'Remembrance'.
Hardy's 'The Ballad of Love's Skeleton', iii. 269, 'The Announcement', ii. 205, with more triple rhythms.

$a^4b^3c^4b^3$
d.r.

Ballad measure. Variously considered a resolved septenary or fourteener, or a clipped form of long metre ($a^4b^4a^4b^4$). In Tate and Brady, Hardy marked and annotated 1, 13, 23, 34, 41, 42, 44, 78, 81, 84, 90, 102, 119, 121, 133, 143. Hymn and ballad forms are generally but not strictly distinguishable in that common hymn metre tends to follow a stricter iambic rhythm in an $a^4b^3a^4b^3$ pattern while ballads tend to use an $a^4b^3c^4b^3$ pattern as in 'The Outlandish Knight' quoted by Hardy in the *Life*, 20. In his early 'Poetical Matter' notebook, Hardy wrote: 'ballad metre . . . [Rhyme, only 2nd and 4th lines]' (Millgate, *Biography*, 89). Historically, each form influenced the other, and in various ways derived from Old English accentual poetry and from Latin hymnody when the hymn writers began to emphasize stress rather than quantity and began to use more terminal rhymes. *Sequences from the Sarum Missal*, which Hardy used, contains several examples of these Latin hymns; see below $a^4a^4b^4c^4c^4b^4$. Hardy's investigation of 'Latin hymns at the British Museum' in 1900 (*Life*, 306) and his imitation of some *Sarum Missal* forms may represent his own interest in this history.

Hardy's 'The Casterbridge Captains', i. 63, 'The Widow Betrothed', i. 177, 'The Division', i. 270, 'An Upbraiding', ii. 282, 'On Stinsford Hill at Midnight', ii. 365, 'The Carrier', iii. 12, 'The Pair He saw Pass', iii. 75, 'A Practical Woman', iii. 219, 'On J.M.B.', iii. 305. Two others, 'Her Immortality', i. 180 and 'The Third Kissing-gate', iii. 246, follow a strict iambic pattern like a hymn. In trisyllabic rhythm: 'At the Draper's', ii. 146, 'Memory and I', i. 226 (alternate stanzas); see also $a^4b^3c^4b^3d^4e^3f^4e^3$.

$a^4b^4c^4b^2$
d.r.

Keats's 'La Belle Dame Sans Merci' (Palgrave, 193—Hardy adds double quotes in his copy to the warrior's cry in the 10th stanza); De La Mare's 'The Glimpse'; William Sharp's 'On a Nightingale in April', *Oxford Book of Victorian Verse*, in more trisyllabic rhythm. Hardy's 'At a House in Hampstead', ii. 340 with more trisyllabic rhythms.

$a^4b^4c^4b^3$
d.r.

Wordsworth's 'Lines Written in Early Spring'; Palgrave 168 (Coleridge's 'All Thoughts, All Passions, All Delights'); De La Mare's 'In Vain'; other examples in Schipper, ii. 565.

Hardy's 'A Woman's Fancy', ii. 341. With more trisyllabic rhythms: 'The Rival', ii. 166.

a⁴b⁴c⁴b⁴
t.f.

'Early One Morning' in Hullah, 88, inscribed by Hardy. This poem changes from t.r. in the first 3 lines to t.f. in the last line of each stanza.
Hardy's 'The Going of the Battery', i. 119.

a³b²c³d²
with all the
d lines rhyming
d.t.r.

Ghazal Stanza. Barnes's 'Woak Hill' part of which is quoted in *Far from the Madding Crowd*, LVI, p. 453. Also see *Personal Writings*, 84, 105.
Hardy's 'My Cicely', i. 67, 'The Mother Mourns', i. 144, 'The Flirt's Tragedy', i. 258; *Dynasts*, 5. Barnes's *d* rhyme in this and in his other examples of the ghazal stanza ('The Knoll', etc.) consists in the same word or phrase, which is thus like a traditional English refrain. Hardy's *d* rhymes are different words and thus carry a more experimental 'Persian' air. Nineteenth-century interest in Persian literature, inspired by the researches of Sir William Jones and given notable impetus by the *Rubáiyát* in 1859, became surprisingly strong after Fitzgerald's death in 1883 (the year Hardy wrote the note that led to 'The Mother Mourns'—*Life*, 163). The *fin de siècle* saw the founding of the Omar Khayyám club and an extraordinary outpouring of articles on and imitations of Persian literature. Hafiz and his ghazals were perhaps even more celebrated than Khayyám: see John Yohannan, *Persian Poetry in England and America* (Delmar, NY, 1977). Hardy had quoted Gosse's *Firdausi in Exile* in *The Woodlanders* and dined at the Omar Khayyám club, of which Gosse was president, in 1895 (*Life*, 268). Hardy's four ghazals may have been written in response to this Persian revival.

Quintets

$a^4a^4a^4b^4C^2$
d.r.

Aldrich's 'Batuschka' in *The Sisters' Tragedy* (London, 1891), 31. Hardy's copy (with pages 185–92 uncut) in the Dorset County Museum is signed by him and the poem 'Apparition' marked.

Hardy's 'We be the King's Men', in *Dynasts*, 9.

$a^3a^3b^2b^2a^3$
d.r.

In rising rhythm: Herrick's 'Her Eyes the Glow-worme Lend Thee', in Henley; Moore's 'O Where's the Slave So Lovely', in Hullah, 278; Moore's 'Oh, The Sight Entrancing' (from *Irish Melodies*) in Linton combines three such stanzas into one.

Hardy's 'To Outer Nature', i. 80, in falling rhythm however.

$a^2b^2a^2a^2b^2$
d.r.

Newman's 'The red sun is gone' from *Verses On Various Occasions* (London, 1868); Hardy's copy is in the Dorset County Museum, along with Hardy's copy of Newman's *Grammar of Assent* (London: 1870) and the *Apologia* (London: 1879); for Newman, see also $a^4b^4a^4b^2$, $a^2a^2b^4c^2c^2b^4$, $a^5b^2a^5b^2c^5c^6$; and *Literary Notebooks*, i. 5–7, with Björk's notes.

Hardy's 'Looking Across', ii. 243.

$a^4b^3a^4a^4b^3$
d.r.

Common tail-rhyme stanza, as in *Hymns* (1889), 'Lo! Now the Time Accepted Peals', 492; 'To You Who Live at Home in Ease', in Hullah, 66; Monk Lewis's 'Alonzo the Brave', one of Hardy's favourites (Millgate, *Career*, 289), in triple rhythm. For many others, see Schipper, ii. 578.

Hardy's 'New Year Eve', i. 334; also 'God's Education', i. 335, 'His Country', ii. 290, 'At Middle-Field Gate in February', ii. 220, 'The Sun's Last Look on the Country Girl', ii. 474, these last two in mixed rhythms.

$a^4b^4a^4a^4b^2$
d.r.

Grant Allen's 'Pessimist' from *The Lower Slopes*. Hardy owned a copy, inscribed to him by Allen, and said it 'is a volume I prize very much as coming from you' (*Collected Letters*, ii. 58). Hardy met Allen on various occasions at the intellectual gatherings at Aldeburgh. See *Collected Letters*, i. 277, Edward Clodd's *Memories* (London, 1900), and his *Grant Allen: A Memoir* (London, 1900). Also see above $a^5b^5a^5b^2$.

Hardy's 'Going and Staying', ii. 338; in d.f., 'The Peace-Offering', ii. 201.

$a^4b^4a^4a^4b^3$
d.t.r.

Charles Sackville's 'On the Countess of Dorchester'. The *Oxford Book of English Verse* has one example of Sackville's work. Hardy's 'Architectural Masks', i. 199, 'Faintheart in a Railway Train', ii. 329, the first stanza of 'A Thunderstorm in Town', ii. 18; some stanzas in 'The Bride-Night Fire', i. 93 in more trisyllabic rhythm.

$a^5b^5a^5a^5b^5$
d.r.

Herbert's 'The World'; Rosamund Watson's 'In a London Garden', marked by Hardy in his copy of Watson's *Poems* (London, 1912—Hardy's copy, DCM); Mary Coleridge's 'Some in a Child', 198.

Hardy's ' "And There Was a Great Calm" ', ii. 355.

$a^3b^3a^3b^3b^3$
d.t.r.

Browning's 'On the First of the Feasts of Feasts', part of the *Epilogue* to *Dramatis Personae*.

Hardy's 'A Merrymaking in Question', ii. 203. In trisyllabic rhythm: 'The Marble Tablet', ii. 433.

$a^4b^3a^4b^3b^2$
d.r.

Dante Rossetti's 'The Staff and Scrip'.

Hardy's 'The Absolute Explains', iii. 68.

a⁴b⁴a⁴b⁴b²
d.r.

Swinburne's 'Félise' marked by Hardy in his copy of *Poems and Ballads* i; *Hymns* (1899), 629, 'Lord I Hear of Showers of Blessing', marked and listed in the back by Hardy; Saintsbury, ii. 132 cites an older example of this stanza, 'All day I weepe my wearie woes', from *The Phoenix Nest*, 51.
Hardy's 'At Castle Boterel', ii. 63, 'Old Furniture', ii. 227, though Hardy uses more trisyllabic substitutions.

a⁴b⁴a⁴b⁴b⁴
d.r.

E. B. Browning's 'A View across the Roman Compagna'; Mary Coleridge's 'Our Lady', 'One and All', 'When Mary Thro' the Garden'; De La Mare's 'Keep Innocency'; Monckton Milnes's 'The Treasure Ship', in Linton; Wotton's 'You Meaner Beauties' (Palgrave, 84); Carew's 'When Thou, poor Excommunicate', in the *Oxford Book of English Verse*; Swinburne's 'Aholibah'; Sidney's 'Psalm 4'.
Hardy's ' "I looked back" ', iii. 255.

a⁵b⁵a⁵b⁵b²
d.r.

Sidney's 'Psalm 9'.
Hardy's 'The Prospect', iii. 87.

a⁵b⁵a⁵b⁵b⁵
d.r.

Donne's 'Hymn to God my God, in My Sickness'; Mary Coleridge's 'Not as I am'. See other examples in Schipper, ii. 552.
Hardy's 'Alike and Unlike', iii. 108. In more varied rhythm: 'Last Week in October', iii. 15.

a⁴b⁴b⁴a⁴a⁴
d.t.r.

John Masefield's 'The Harper's Song' from *Ballads and Poems* (London, 1910—Hardy's copy, DCM). Hardy told Masefield in 1911: 'I bought your "Ballads & Poems", & have marked several in it that I like' (*Collected Letters*, iv. 189); Sidney's 'Psalm 28'.
Hardy's 'The Recalcitrants', ii. 107; ' "Sacred to the Memory" ' (d.r.), ii. 452.

a⁴b⁴b⁴a⁴b⁴
d.r.?

Barnaby Barnes's 'Lovely Maya, Hermes's Mother' from *The Shakespeare Anthology*—though Hardy's reading the poem here postdates 'In the Vaulted Way'.
Hardy's 'In the Vaulted Way', i. 275, 'Plena Timoris', iii. 53.

a⁴b⁴b⁴a⁴C²
d.r.

Heine's 'O List to this Spring Time's Terrible Test' on a page (139) in Bowring marked by Hardy. (See above, a³b³a³b³ d.r.) Heine's poem is more trisyllabic than Hardy's.
Hardy's 'Four in the Morning', iii. 22.

a³b²c³b²b⁵
d.r.

Elizabeth Foote's 'The Inter-veil', *Century Magazine* (March 1864), quoted by Hardy in *Literary Notebooks*, ii. 158–9.
Hardy's 'A Wife and Another', i. 318.

a⁴b³c⁴c⁴b³
d.r.

Rosamund Watson's 'Ballad of the Bird-Bride' from *The Bird-Bride* (London, 1889—Hardy's copy, Maggs catalogue).
Keble's 'Fourth Sunday in Lent', marked 'Stinsford/62' by Hardy.
Hardy's ' "I rose up as my custom is" ', ii. 94.

Sestets

a⁴a⁴b⁴b⁴c⁴c⁴
d.r.

Nine examples marked by Hardy in *Hymns* (1889), including 49, 428, and 606, 'O Father, in Whose Great Design', which are all listed by Hardy in the back; Palgrave, 124 (Collins's 'How Sleep the Brave'), 163 (Sewell's 'The Dying Man in His Garden'). See Schipper, ii. 609 for many examples.
Hardy's 'A Necessitarian's Epitaph', iii. 228; *Dynasts*, 429; portions of other poems ('The Three Tall Men', iii. 187, 'On the Death-Bed', ii. 147, 'Last Words to a Dumb Friend', ii. 435, 'Under the Waterfall', ii. 45, 'An August Midnight', i. 184); fourth stanza of 'A Nightmare, and the Next Thing', iii. 202.

$a^4a^4b^4b^4C^4D^4$
d.r., but d.f. in last
two lines repeated
as refrain

'Richard of Taunton Dean; or Dumble dum deary', an old song cited in Hickson, 51.
Hardy's 'The Sergeant's Song', i. 23, which also repeats the second couplet as a refrain.

$a^4a^4b^4c^3b^4c^3$
$a^4a^3b^4c^3b^4c^3$
d.r.

'Young Hal', called 'a favorite song of Haydn's' in Hardy's 1799 Music Book in *Original Manuscripts*, in the first scheme.
Hardy's 'The Dream Is—Which?' ii. 427, in the second scheme.

$a^2a^2b^3c^2c^2b^3$
d.r.

Common tail-rhyme scheme as in Herrick's 'A Ring Presented to Julia' and others; also Henry Carey's 'I am in truth a country youth' in Hullah, 98; Miller Price's Song in Hardy's 'Words of Old County Songs of 1820 onwards' in *Original Manuscripts*, though the rhythm here is more trisyllabic than in Hardy's poem. For older examples of these and other tail-rhyme forms, see C. Strong, 'History and Relations of the Tail-Rhyme Strophe in Latin, French, and English', *PMLA* 22 (1907), 371–420.
Hardy's 'To a Sea-cliff', iii. 113.

$a^2a^2b^4c^2c^2b^4$
d.r.

Newman's 'Days herald bird', the Lauds-Tuesday poem from *Verses*; Wyatt's 'Consent at Last', in Guest, 587; Darley's 'Oh, Sweet to Rove', 'Hither! hither!', and others from *Sylvia, or, the May Queen* (93, 129, 163) and *Miscellaneous Poems* (437). John Dart's Song 'She Hears Me Not' in Hardy's 'County Songs of 1820 Onwards', except that the song has a trimeter sixth line and a strict iambic rhythm.
Hardy's 'Self-Unconscious', ii. 40, 'Outside the Casement', ii. 443.

$a^2a^2b^5c^2c^2b^5$
d.r.

Keats's Roundelay 'O Sorrow' in Linton and in Henley.
Hardy's 'The Prophetess', iii. 170.

$a^3a^3b^3c^3c^3b^3$
d.r.

Hymns (1889), 303 marked by Hardy; Charles Mackay's 'The Holly Bough' and Richard Dixon's 'The feathers of the willow', in the *Oxford Book of Victorian Verse*, 301, 509.
Hardy's 'Transformations', ii. 211.

$a^3a^3b^5c^3c^3b^5$
d.r.

Keble's 'Third Sunday in Advent' (marked 'S/61' by Hardy) and 'Second Sunday After Easter' (marked 'Kilburn' by Hardy); Collins's 'Ode to Simplicity', *Oxford Book of English Verse* (though Hardy's reading Collins here may postdate 'Sapphic Fragment'); Ingelow's 'The Nightingale Heard by the Unsatisfied Heart'; Sidney's 'Psalm 26'.
Hardy's 'Sapphic Fragment', i. 222.

$a^4a^4b^2c^4c^4b^2$
d.r.

Jonson's 'Rime, the rack of finest Wits', in Guest, 660; Locker's 'The Housemaid' in *London Lyrics* and 'To My Grandmother' in the *Oxford Book of Victorian Verse*. See *Life*, 133 for Hardy's enthusiasm for Locker, and also see $a^3b^2a^3b^2$, $a^2b^2a^2b^2c^2d^2c^2d^2$, $a^4b^2a^4b^2c^4d^2c^4d^2$. Saintsbury also likes Locker (iii. 268).
Hardy's 'The Chimes', ii. 215.

$a^4a^4b^2c^4c^4b^2$
d.f.

Rosamund Marriott Watson's 'A Portrait' from *The Bird-Bride*: for his admiration for Watson ('Graham Tomson'), see Millgate, *Career*, 302, 407; *Collected Letters*, i. 249, ii. 24; Gittings, *Older Hardy*, 64–5; see also above, $a^5b^5a^5a^5b^5$ and $a^4b^3c^4c^4b^3$. Fletcher's 'Roses, their sharp spines being gone' in *The Shakespeare Anthology*.
Hardy's 'The Sigh', i. 227 and 'To a Well-Named Dwelling', ii. 453. If 'The Sigh' is from the 1860s, however, it predates Watson.

a⁴a⁴b³c⁴c⁴b³
d.r.

Common tail-rhyme (parodied in 'Sir Thopas'). *Hymns* (1889), 276 (labelled 'Dorset' by Hardy), 139, 314, both these marked by Hardy. 'Ye Cheerful Virgins', in Hardy's 1799 Music Book, p. 66. Keble's 'First Sunday after Easter' (marked 'K/62' by Hardy) and others by Keble also marked by Hardy; Palgrave, 120 (Gray's 'On a Favourite Cat'), 179 (Wordsworth's 'Three Years She Grew in Sun and Shower'); Lodges's 'The Barginet of Antimachus', in *The Shakespeare Anthology*; Sidney's 'Psalm 32'; Heine's 'What Drives Thee On' on a page (186) marked by Hardy in Bowring (see above, a³b³a³b³ d.r.).
Hardy's 'They are the shipped battalions sent', in *The Dynasts*, 197; 'We do not know', in *The Queen of Cornwall*, ii, p. 10, with more trisyllabic rhythms.

a⁴a⁴b⁴c⁴c⁴b⁴
d.f.

Adam of Saint Victor's 'Heri Mundus Exultavit'; also other such hymns from *Sequences from the Sarum Missal* owned by Hardy (see *Life*, 306; Purdy, 237, 240); Heine's 'Presentiment' in the Bowring translation; Palgrave, 26 (Shakespeare's 'O Mistress mine') d.f. after the first two lines. Hardy's 'Sine Prole', iii. 30. Using different rhythms and line lengths, Hardy wrote thirteen other poems in this common tail-rhyme scheme. See Hickson, 101.

a³b³a³b³a³b³
d.r.

Drayton's 'An Ode Written in the Peake'. Hardy cites *Polyolbion* by Drayton in *Jude* (see above, Chapter 2) and in *Collected Letters*, iii. 133.
Hardy's 'Heredity', ii. 166, 'The Master and the Leaves', ii. 434.

a⁴b²a⁴b²a⁴b²
d.r.

Louise Moulton's 'The Sun is Low' from *At the Wind's Will* (Boston, 1900—Hardy's copy, Holmes Listing), first and last stanzas. Hardy said of this volume: 'some of them I thought very beautiful' (*Collected Letters*, ii. 266); for Moulton, see also $a^4b^3c^4b^2$.

Hardy's ' "I sometimes think" ', ii. 332, 'The Seasons of Her Year', i. 195.

$a^4b^3a^4b^3a^4b^3$
d.r.

Extended 'common' hymnal metre, as used in Byron's 'The Days are Done'.

Hardy's 'The Ivy-Wife', i. 75, 'Drummer Hodge', i. 122, 'The Two Soldiers', ii. 113, 'First Sight of Her and After', ii. 165, ' "The Wind blew words" ', ii. 181, 'The Lament of the Looking-Glass', ii. 456, 'The Impercipient', i. 87, and 'Shut out that Moon', i. 265 (last stanza).

$a^4b^4a^4b^4a^4b^4$
d.r.

Extended 'long' hymnal metre, as used by Byron in 'She Walks in Beauty' (Palgrave, 173); *Hymns* (1889), 67 ('Word Supreme'), 179 ('To the Name of Our Salvation'), 309 ('Now, my Tongue, the Mystery Telling'), and 385 ('God the Father'), all marked by Hardy; Sidney's 'Psalm 36'.

Hardy's 'A Dream Question', i. 316.

$a^4b^4a^4b^4b^4a^2$
d.r.

John Freeman's 'The Winds' from *Memories of Childhood* in *Poems New and Old* (London, 1920). Citing Freeman, among others, Hardy said: 'There is a lot of good work being done' (*Talks*, 6).

Hardy's 'The Rift', ii. 395.

$a^3b^3a^3b^3c^3c^2$
$a^3b^2a^3b^3c^3c^2$
$a^3b^3a^3b^4c^3c^2$
d.r.

George Meredith's 'Ameryl's Songs—1' in the first scheme (from *The Shaving of Shagpot* in *Songs from the Novelists*, 139): On 16 August 1885 Hardy thanked the publishers for sending him this volume and said he found it 'highly interesting' (*Collected Letters*, i. 135); see also $a^4a^4b^4b^4c^2c^2c^4$, $a^4b^2a^4b^2c^4c^4c^4a^2$.

Hardy's 'Looking at a Picture on an Anniversary', ii. 283 in the second scheme; 'A Self-glamourer', iii. 208 in the third scheme. Meredith's poem is more trisyllabic than Hardy's and is written in headless lines.

a³b³a³b³c⁴c⁴
d.r.

Hymns (1889), 319 'Author of Life Divine' marked by Hardy; Campion's 'Some, the quick eye commends', in *The Shakespeare Anthology*; also see Schipper, ii. 648.
Hardy's 'Middle-Age Enthusiasms', i. 82.

a⁴b³a⁴b³c⁴c⁴
d.f.

Shelley's 'Invocation' (Palgrave, 226).
Hardy's 'Mad Judy', i. 189.

a⁴b⁴a⁴b⁴c⁴c⁴
d.r.

Common sestet. Eleven examples are marked by Hardy in *Hymns* (1889), including 192, 'O Love, Who Formedst me to Wear', marked 'Rg'; 345, 'O Light, Whose Beams Illumine All', marked 'E.L.H.' and listed in the back with other examples; 204, 'O Quickly come, dread Judge of all'; 401 'Now the labourer's task is o'er'; and 600 'Thou hidden love of God, whose height'.
Hardy's 'Your Last Drive', ii. 48.

a⁴b⁴a⁴b⁴c⁴c⁴
t.d.r.

Schiller's 'The Diver', which Hardy marked in his edition of *Poems and Ballads* (London, 1887), trans. Bulwer-Lytton (DCM); many English examples in Schipper, ii. 616.
Used by Hardy in many of the 'Satires of Circumstance': 'At a Watering-Place', ii. 142, 'By Her Aunt's Grave', ii. 141, 'Outside the Window', ii. 143, 'At Tea', ii. 140, 'In the Nuptial Chamber', ii. 145, 'In the Restaurant', ii. 146; and in one stanza in five others: 'In the Cemetery', ii. 143, 'In the Room of the Bride-Elect', ii. 142, 'In Church', ii. 140, 'At the Altar-Rail', ii. 145, and 'Over the Coffin', ii. 148. Also 'An August Midnight' (first stanza), i. 184.

$a^5b^2a^5b^2c^5c^6$
d.r.

Newman's 'St. Gregory Nazianzen', from *Verses*. 'Lead, Kindly Light' is similar but ends with a pentameter.
Hardy's 'Concerning Agnes', iii. 215.

$a^5b^5a^5b^5c^5c^5$
d.r.

Heroic Sestet. *Hymns* (1889), 322 marked by Hardy. Palgrave, 41 (Vere's 'If Women Could Be Fair'), 240 (Wordsworth's 'To A Skylark'). Campion's 'When Thou Must Home' marked by Hardy in his copy of Henley; Newbolt's 'Youth' marked by Hardy in his copy of *Songs of Memory and Hope* (London, 1909–DCM).
Hardy's 'The Chimes Play "Life's a Bumper!" ', ii. 375.

$a^5b^5c^5a^5b^5c^5$
d.r.

Italian sestet.
Hardy's 'Evening Shadows', iii. 186.

$a^2b^2c^2b^2d^2b^2$
d.r.

De La Mare's 'Fear' in Hardy's copy of *Poems* (London, 1906); other citations of De La Mare in this Appendix are under: $a^4b^3c^4b^2$, $a^4b^4c^4b^2$, $a^4b^4c^4b^3$, $a^4b^4a^4b^4b^4$.
Hardy's 'The Robin', ii. 264.

$a^4b^3c^4b^3d^4b^3$
d.r.

Extended ballad stanza as in Rossetti's 'The Blessed Damozel' and 'The Card-Dealer', the latter in Linton.
Hardy's 'The Single Witness', iii. 257, 'Shut out that Moon', i. 265 (first stanza), ' "You on the tower" ', ii. 230 (second stanza).

$a^4b^3c^4c^4a^4b^3$
d.r.

Catullus' 'Of All the Many Loved by Me', translated by Lamb, in Hardy's copy of Catullus and Tibullus, *Poems* (London, 1887—Colby College Library).
Hardy's 'The Unborn', i. 343.

$a^4b^3c^4c^4c^4b^3$
d.t.r.

Lawrence Hope's 'Sampan Song' from *The Garden of Kama and Other Love Lyrics from India*; on Hope, see *Life*, 322, *Personal Writings*, 256, *Collected Letters*, iii. 142, Taylor, *Hardy's Poetry*, xv, and Carroll Wilson (94), who cites Hardy's inscribed copy of the third edition (London, 1903). Hardy's ' "Ah, are you digging on my grave?" ', ii. 38, though Hardy's poem is less trisyllabic than Hope's.

Septets

$a^4a^4b^4b^4c^2c^2c^4$
t.r.

George Whyte-Melville's 'Lord Goring's Song' (from *Holmby House*) in *Songs from the Novelists*, 112; see also $a^3b^3a^3b^3c^3c^2$.
Hardy's 'The Dark-Eyed Gentleman', i. 295.

$a^4a^4b^4b^4c^4c^4c^4$
d.r.

Shelley's 'I loved—alas! our life is love' in *Tasso*; other examples in Schipper, ii. 624. Hardy's 'After the Death of a Friend', iii. 196. In duple falling rhythm: 'On a Fine Morning', i. 165.

$a^2a^2b^2c^2c^2c^2b^2$

'A Loyal Song' in Hardy's 1799 Music Book in *Original Manuscripts*, in d.r. rhythm.
Hardy's 'The Maid of Keinton Mandeville', ii. 326 (i.e. Í hear that máiden still / Óf Keinton Mándeville', etc.), in d.f. rhythm, which the dipody easily expands into tetrameter. The poem refers to Henry Bishop's 'Should He Upbraid'.

$a^5b^5a^5b^5b^5c^5c^5$
d.r.

'Rime Royal'.
The rhyme scheme is used by Hardy in 'A Parting-scene', iii. 132 $(a^3b^3a^3b^3b^5c^5c^5)$; 'I see red smears', in *Dynasts*, 299 (all pentameter lines except for the fifth trimeter line).

$a^4b^3a^4b^3c^2c^2b^3$

Ingelow's 'A Sea Song' (t.r.).
Hardy's 'The Church-Builder', i. 210 (d.r.).

$a^4b^3a^4b^3c^4c^4b^3$
d.r.

Thomas D'Arcy McGee's 'The Penitent Raven' in Linton; Cowper's 'To Babylon's Proud Waters' (Psalm 137). Schipper's examples, ii. 661, use feminine rhymes.
Hardy's 'The Dance at the Phoenix', i. 57.

$a^4b^4a^4b^3c^2c^2b^3$
d.r.

Hardy's ' "O I won't lead a homely life" ', ii. 424 has title subscript: 'To an old air'.

$a^4b^4a^4b^4c^2c^2B^3$
$a^4b^4a^4b^4c^2c^2b^4$
d.r.

Chatterton's 'Song from *Aella*', stanza 1, in the first scheme, in Ward 419 and *Oxford Book Of English Verse* 479. *Aella* is quoted in *The Woodlanders*, ch. 25, and much marked in Hardy's 1875 edition of Chatterton's poems (Dorset County Museum).
Hardy's 'The Going', ii. 47 (stanzas 1, 3, 5) in the second scheme.

$A^4B^4A^4C^3A^4d^4C^7$
$A^4B^4A^4c^3A^4a^4C^7$
d.t.r.

'Bonnie Susan Cleland' from *Ballad Minstrelsy*, 78, in the first scheme, with the *A* lines identical in each stanza, and the *B* lines and *C* lines repeated with slight variation throughout the poem. Hardy put a mark beside this poem both in the text and the index of his copy of *Ballad Minstrelsy* (one of 3 so marked).
Hardy's 'The Change', ii. 190 in the second scheme, with the *A* lines identical in each stanza, and the *B* line repeated with slight variation throughout the poem.

Octaves

$a^2a^2a^2b^2a^2a^2a^2b^2$
d.r.

Bottom's song, 'The raging rocks', *A Midsummer-Night's Dream*, I. ii. 33; Thomas Feille's 'The Controversy' and Anthony Wydville's 'Somewhat musing' in *The Dunbar Anthology*, 193, 180; Alexander Montgomery's 'Aubade' in Henley.
Hardy's ' "I need not go" ', i. 174; 'Meditations on A Holiday', ii. 383 (subtitled 'A New Theme to an Old Folk-Measure') in duple falling rhythm.

$a^4a^4a^4b^3c^4c^4c^4b^3$
d.t.r.

'The Loud Tattoo' in *Songs of the Late Charles Dibdin*, ed. T. Dibdin, in Hardy's copy; for Dibdin, see also $a^3b^3a^3b^3c^3d^3c^3d^3$; 'Oh! Could We Do With This World' and 'O Stay Sweet Warbling Wood-Lark' in Hullah, 146, 195 (both d.t.r.). There are many more strictly iambic examples: Wordsworth's 'The Green Linnet' (Palgrave, 242); Keble's 'St. Stephen's Day' and 'Fourteenth Sunday After Trinity'; Barnes's 'Mëaken up a Miff' (in Hardy's selection); Haynes Bayly's 'Sigh Not for Summer Flowers'.

Hardy's 'The Mongrel', iii. 214 is heavily trisyllabic.

$a^4a^4a^4b^4c^4c^4c^4b^4$
d.f.

Adam Saint Victor's 'Officium Beatae Mariae' and other examples from *Sequences from the Sarum Missal*: see above $a^4a^4b^4c^4c^4b^4$ d.f.

Hardy's 'Genitrix Laesa', iii. 88. Six other Hardy poems, using different line types, are written in this extended tail-rhyme octave: see Hickson, 105.

$a^2a^2b^2a^2c^2c^2c^2a^2$
d.t.f.

Hardy's 'The Colour', ii. 479 has subscript: 'The following lines are partly original, partly remembered from a Wessex folk-rhyme'. Bailey, 496–7, quotes the rhyme, 'Jinny Jones', which Hardy has modified.

$a^4a^4b^4b^4c^4c^4d^4d^4$
d.r.

Extended hymnal form. *Hymns* (1887), 127, 'At the Lamb's High Feast'; 131, 'Christ the Lord is Risen To-day'; 544, 'Praise the Lord'; 610, 'Safely, Safely Gather'd In', all marked by Hardy. Barnes's 'The Blackbird' and others. Marvell's 'The Garden' (Palgrave, 111).

Hardy's 'Faithful Wilson', iii. 233. 'On a Discovered Curl of Hair' (first stanza), ii. 449. 'In a London Flat', ii. 474 follows a similar scheme but in triple rising rhythm.

$a^5a^5b^5b^5c^5c^5d^5d^4$
$a^5a^5b^5b^5c^5c^5d^5d^3$
d.r.

Matthew Prior's 'Considerations on Part of the Eighty Eighth Psalm' in the first scheme; for Hardy on Prior, see *Personal Writings*, 222.

Hardy's 'The Harbour Bridge', iii. 92, in the second scheme.

$A^4B^4a^4A^4a^4b^4A^4B^4$
d.r.

Triolet; several examples like Bridges's 'When first we met' and Gosse's 'Happy, my Life' in *Ballades and Rondeaus,* ed. Gleeson White. White's 'Introduction' may have challenged Hardy: 'the subtle art needed to acquire the ease that is the charm of a good triolet is generally the result of infinite care. . . . the triolet affords so little space to explain its motif, and within its five lines must tell its story, and also carry the three other repeated ones easily' (lxx).

Hardy's ' "How great my grief" ', i. 173, 'The Coquette, and After', i. 175, 'At a Hasty Wedding', i. 179, 'Birds at Winter Nightfall', i. 185, 'The Puzzled Game-Birds', i. 185, 'Winter in Durnover Field', i. 186.

$a^4b^3a^4b^3a^4b^3a^4b^3$
d.f.

Doubled hymn stanza, in falling rhythm. Hardy's 'Wives in the Sere', i. 182.

$a^5b^5a^5b^5a^5b^5c^5c^5$
d.r.

Ottava rima, as in Palgrave, 268 (Shelley's 'I Dream'd That as I Wander'd').

Hardy, last stanza of 'A Singer Asleep', ii. 31, except that the last line is a trimeter; the last stanza of 'Over the Coffin', ii. 148 is in this rhyme scheme with tetrameter lines.

$a^4b^4a^4b^4b^4c^4b^4c^4$
d.r.

Burns's 'Mary Morison' (Palgrave, 148); Byron's 'If That High World'; for other examples, see Schipper, ii. 626; this is Chaucer's 'Monk's Tale' rhyme scheme.

Hardy's 'She at His Funeral', i. 14; also 'Her Initials', i. 15, 'On an Invitation to the United States', i. 142, 'Catullus: XXXI', i. 221, 'A Maiden's Pledge', ii. 380.

$a^3b^2a^3b^2c^3c^3c^3b^2$
d.f.

Burns's song, 'Phillis the Fair'. Hardy's 'In a Wood', i. 83.

$a^2b^2a^2b^2c^2d^2c^2d^2$
d.r.

Hymns (1889), 167 ('O Worship the King') and 308 ('O Praise Ye the Lord') both

marked by Hardy; Byron's 'When We Two Parted' (Palgrave, 190), praised by Saintsbury, iii. 96; 'Over the Mountains' (Palgrave, 80) marked in the index by Hardy; Palgrave 234 (Scott's 'Coronach' d.t.r.); Locker's 'Geraldine Green'.
Hardy's 'Evelyn G. of Christminster', ii. 393.

$a^3b^2a^3b^2c^3d^2c^3d^2$ d.r.

Bryan Waller Procter's 'The Poet To His Wife' in Linton.
Hardy's 'The Rejected Member's Wife', i. 346 in more varied rhythm; also 'At an Inn', i. 89, 'At the Word "Farewell"', ii. 364 in trisyllabic rhythm. '"Let me believe"', iii. 18, with the dimeter lines in falling rhythm.

$a^3b^3a^3b^3c^3d^3c^3d^2$ d.t.r.

Swinburne's 'Dolores' marked by Hardy in his 1873 *Poems and Ballads* I (Swinburne uses more trisyllabic rhythms and feminine endings than Hardy); Untermeyer's 'Haunted' in *Challenge* (New York, 1914) with no feminine endings and with strict iambic rhythm: in 1923 Hardy said: 'You have some mighty promising poets in America. Who is Louis Untermeyer? I'm very fond of his verse' (Brennecke, *Life*, 6).
Hardy's 'The Strange House', ii. 346.

$a^3b^3a^3b^3c^3d^3c^3d^3$ d.r.

Hymns (1889), 111 ('O Sacred Head'), 226 ('The World is Very Evil'), 337 ('There's a Friend'), 358 ('From Greenland's Icy Mountains'), 632 ('Redeem'd, Restored, Forgiven')—all marked and the last listed in the back by Hardy, and all with alternate feminine rhymes, as in Hardy; Sidney's 'Who hath his fancy pleased', in Henley, 'The Greenwich Pensioner' and 'Poor Ship-wreck'd Tar', in Dibdin.
Hardy's 'In a Eweleaze near Weatherbury', i. 92, 'Mute Opinion', i. 162; 'The Farm-Woman's Winter', i. 262,

'"Known had I"', iii. 128, 'Christmastide', iii. 179. In duple falling rhythm, 'A Death-Day Recalled', ii. 61; in triple falling rhythm, 'The Old Gown', ii. 351.

$a^3b^3a^3b^3c^4d^3c^4d^3$ d.r.

Two different hymn quatrains combined. Used in 'The Fairies Farewell', in Percy's *Reliques*; for other examples, see Schipper, ii. 669. Moore's 'To-day Dearest' is close but more rhythmically complex.

Hardy's 'Her Song', ii. 343. In all these examples, feminine rhymes alternate in the first quatrain.

$a^4b^2a^4b^2c^4d^2c^4d^2$ d.r.

Locker's 'A Rhyme of One' (see above $a^4a^4b^2c^4c^4b^2$ d.r.); Leigh Hunt's 'To His Piano-Forte' in Linton; Landor's 'Songlets' in Henley; Lord Houghton's 'Shadows', part VI ('They tell me I have Won Thy Love'), marked by Hardy in his copy of Houghton (see $a^5b^2a^5b^2$).

Hardy's '"A Gentleman's Second-hand Suit"', iii. 221.

$a^4b^3a^4b^3c^4d^3c^4d^3$ d.r.

Combined hymn stanzas. Barnes's 'The Wife a-Lost', one of Hardy's favourites (*Personal Writings*, 84, 96–7); Palgrave, 99 (Lovelace's 'To Althea from Prison') marked in the index by Hardy; see others in Schipper, ii. 525–6; Byron's 'All is Vanity' and others (Schipper, ii. 526) use alternate feminine rhymes.

Hardy's 'The Inconsistent', i. 171, 'The Darkling Thrush', i. 187, 'The Rover Come Home', iii. 127 (third stanza), *Queen of Cornwall* XI, p. 40, *Dynasts*, 197, 389, 392; also 'Retty's Phases', iii. 110, 'The Haunter', ii. 55 (last two stanzas), the last two examples in headless lines; '"When wearily we shrink away"', iii. 283, like Byron's example, uses alternate feminine rhymes.

$a^5b^5a^5b^5c^5d^5c^5d^5$
d.r.

Doubled heroic quatrain.
Hardy's 'The Children and Sir Nameless', ii. 399.

$a^3B^2a^3c^2a^3a^3a^3c^2$
d.r.

Hymn, 'Remember Adam's Fall', quoted in *Under the Greenwood Tree*, I. 4, p. 27.
Hardy's 'Dead "Wessex" the Dog to the Household', iii. 258.

$a^2b^2b^2c^2d^2d^2d^2c^2$
d.f.

Hardy's 'Timing Her', ii. 178, with the subscript 'Written to an old folk-tune'.

$a^4b^4c^4b^4d^2b^4e^2b^4$
$a^4b^4c^4b^4d^4b^4e^2b^4$
d.f.

Horace Moule's 'Ave Caesar' (four middle stanzas) in the first scheme, praised by Hardy as a 'fine poem' and reprinted by him in the *London Mercury*, 6 (Oct. 1922), 631–2. A newspaper clipping of the poem is pasted in the back of Hardy's *Golden Treasury* and dated '1862' by him.
Hardy's 'Cry of the Homeless', ii. 296 in the second scheme.

$a^4b^3c^4b^3d^3e^3d^3e^3$
$a^4b^3c^4b^3d^4e^4d^4e^4$
d.r.

This octave combines a ballad and common hymn quatrain. *Hymns* (1889), 'O Paradise', 234, marked by Hardy, in the first scheme.
Hardy's 'The Whitewashed Wall', ii. 470 in the second scheme.

$a^2b^2c^2b^2d^2e^2f^2e^2$
d.r.

Thomas Haynes Bayly's 'Welcome Me Home' is in d.f. rhythm. Also see $a^4a^4a^4b^3c^4c^4c^4b^3$ above, $a^3b^3c^3b^3d^3e^3f^3e^3$ below. On Bayly, see also *Life*, 14.
Hardy's 'The Catching Ballet of the Wedding Clothes', iii. 264.

$a^3b^3c^3b^3d^3e^3f^3e^3$
d.r.

Hymns (1889), 271 ('O Jesus, I have Promised'), 379 ('Now Thank We All Our God'), 406 ('We Sing the Glorious Conquest'), 500 ('O Voice of the Beloved'), 607 ('O Thou Before Whose Presence'), 613 ('Praise to the Heavenly Wisdom'), 621 ('Come sing, ye choirs exultant'), all marked by Hardy; Monckton Milnes's

'The Brook-Side', in Linton and also marked by Hardy in his copy of Milnes; Haynes Bayly's 'My Heart is All Alone'. Hardy's 'The Dolls', ii. 241.

$a^4b^3c^4b^3d^4e^3f^4e^3$
d.r.

Combined ballad quatrains. Palgrave, 151 (Burns's 'Highland Mary'), etc. *Hymns* (1889), 186, 257, 229, 357, 170, 256, 369 this last marked by Hardy, and the last three listed in the back.
Hardy's 'The Rover Come Home', iii. 127 (first stanza); 'The Haunter', ii. 55 (first two stanzas).

Long Stanzas

$a^5b^5a^5b^5b^5c^5b^5c^5c^6$
d.r.

Spenserian stanza.
In the 1860s Hardy 'began turning the Book of Ecclesiastes into Spenserian stanzas, but finding the original unmatchable abandoned the task' (*Life*, 47). The experiment has not survived. In his copy of Thomson, Beattie, *et al.* (signed by Hardy in 1865), he marked the first stanza of Beattie's 'The Minstrel'; the stanza puts an Ecclesiastes theme into a Spenserian stanza.

$a^4b^2a^4b^2c^4c^4c^4c^4a^2$
$a^4b^2a^4b^2c^4c^4c^4c^4b^2$
d.r.

Thomas Lodge's 'Rosalynd's Madrigal' in the first scheme, from *Songs from the Novelists*, 9, also in the *Oxford Book of English Verse* and in *Henley*.
Hardy's 'Weathers', ii. 327 in the second scheme.

$a^4b^3a^4b^3c^4c^4d^3e^4e^4d^3$
d.r.

Byron's 'L'Amitié est l'Amour Sans Ailes', which Hardy marked in his copy of Byron (*Select Works*, London, 1867—DCM) and whose title Hardy translated as: 'Friendship is love without (hid) wings'; Thomas Campbell's 'The Last Man'; Gray's 'Ode on a Distant Prospect of Eton College' (Palgrave, 158) is similar except that its seventh line is a tetrameter. In his copy of Palgrave, next to Gray's last 2 lines, Hardy wrote: 'Cf. Soph: Ædip. 316. *Φεῦθεν' Φρονιν'* etc.' This is about the oldest still used long stanza in English. It was used in 'Iesu, for þi muchele miht' from the medieval Harley Ms. It was printed for example in Thomas Wright (ed.), *Specimens of Lyric Poetry* (London, 1862) with a preface dated 1842. The rhyme scheme was common in 18th-century odes—see Bate, 131; the rhyme was used in 'When Wars Alarm', a song from Hardy's 1799 Music Book, p. 35. Hardy's 'Reminiscences of a Dancing Man', i. 266; the rhyme scheme is used in 'Compassion', iii. 147.

$a^3b^3b^3c^3a^3c^3d^2E^5E^5E^5$ d.r.

Bob and wheel stanza. (The short line is the bob line.) This specific version is unique to Hardy: ' "The Curtains now are drawn" ', ii. 335.

Rondeaux

$a^4a^4b^4b^4a^4a^4a^4b^4C^2$ and $a^4a^4b^4b^4a^4C^2$ d.r. First half of first line used as refrain in *C* lines

Gleeson White, *Ballades and Rondeaus*, cites Voiture and Benserade for reviving the rondeau (xxxi), and quotes Austin Dobson's paraphrase of a rondeau by Voiture, 'You bid me try' (lxii) in this scheme, as is a rondeau like Benserade's 'Metamorphoses d'Ovide en Rondeaux'. This is the rondeau 'in its strict form'. Hardy's 'The Roman Road', i. 320; 'Midnight on Beechen, 187–', iii. 86; 'The Skies Fling Flame', in *Dynasts*, 334: 'Even a rondeau, as correct as one of Benserade', Gosse wrote after reading *The Dynasts* (Gosse, *Life*, 310).

The following examples of Hardy's rondeaux do not follow any of the traditional versions. This is an interesting fact in itself. In his introductory essay on romance forms, Gleeson White said of the rondeau that 'like the sonnet, the perfected form is jealously guarded. The genius which consists in breaking rules is looked upon with suspicion in all these forms, but especially in this one' (lxiii). While Hardy imitated other romance forms with strict conformity, his rondeaux (in the loose sense) represent interesting variations in the history of the form. Saintsbury's description in his *History of English Prosody* may have influenced Hardy: '*Ronde, rondeau, rondel,* 'roundel', etc. . . . though later specialised . . . merely refers to the "coming round" of the refrain; while the repetition may, as obviously, extend to whole lines, to more than one line, or to part of a line worked in according to the taste and fancy of the poet. This repetition, again, may be always at the close, or at the beginning, or at both, or it may work its way through the stanzas in different places, like something settling through clear water at different levels' (iii. 388). Saintsbury points out that the earliest examples apparently derived from 'old simpler forms of song and carol in Southern and Northern France', began to 'crystallise' about the thirteenth century, and became strictly regulated in the fourteenth and fifteenth centuries (iii. 388). In his copy of Saintsbury's edition of *French Lyrics* (d. 1882), Hardy read: 'The essence of the rondeau consisted in the repetition of the first line or part of it at intervals, and by degrees it separated itself into the *triolet,* the *rondel,* and the rondeau proper' (p. x). Edwin Guest, whose *History of English Rhythms* Hardy owned, defined the 'roundle' as a 'short poem of not more than three staves. It admits only two rimes; and repeats the whole or part of the opening couplet as a burthen' (644). Describing ' "When I set out for Lyonnesse" ' in 1924, Hardy said: 'The Poem . . . is one of the many varieties of Roundelay, Roundel, or Rondel. The Rondel in its strict form probably came originally from France' (Bailey, 270). ' "When I set out for Lyonnesse' is in three staves and opens and closes each stanza with the same lines. It uses only two rhymes in each stanza (but like the other examples does not use the same rhymes in all stanzas).

In Hardy's formative years, there was a revival of French forms in English, particularly by Dobson, Lang, Gosse, Swinburne, and Henley. Gosse's *Cornhill* article of July 1877, entitled 'A Plea for Certain Exotic Forms of Verse' was a manifesto for the movement, though his discussion of Benserade was rather wry: 'Benserade, in particular, carried the cultivation of this form of verse to so absurd an excess that he translated the whole *Metamorphoses* of Ovid into rondeaux, and had his monstrous exercise systematically printed at the King's press with elaborate illustrations, at a cost, it is said, of

10,000 francs.' Gleeson White's *Ballades and Rondeaus* was a late definitive anthology of the movement (see Helen Cohen).

a⁴b³b³a⁴A⁴B³ d.r. 3 stanzas	Hardy's ' "When I set out for Lyonnesse" ', ii. 17; ' "As 'twere to-night" ', ii. 348 with more rhythmic variation.
a³b³a³a³b⁵b³a² d.r. 1 stanza	The rhyme scheme, but not this rhythmic scheme, is used in Mary Prolyn's 'Which Way He went', in Gleeson White's *Ballades and Rondeaux*, 151. Hardy's 'Could He But Live for Me' in the *Queen of Cornwall*, VII: first line repeated, in modified form, in last two lines of the stanza.
a⁴a²a⁴a⁴b⁴b⁴A⁴A²A⁴ d.f. 2 stanzas	Hardy's ' "If it's ever spring again" ', ii. 362.

Hardy has many other poems (i.e. ii. 401, 402, 414, 437, etc.) which repeat initial lines as refrain lines, but above I have only cited those which keep to two rhymes per stanza, in the rondeau tradition.

Sonnets (all pentameter lines)

abab bcbc cdcd ee	Spenserian sonnet. Hardy's 'Her Reproach', i. 171, dated 1867: the *d* rhymes are consonance rhymes; see *Life*, 105 on 'inexact rhymes'; see Hardy's story, 'An Imaginative Woman', on 'sonnets in the loosely rhymed Elizabethan fashion'.
abab cdcd efefef efeffe effeef	Shakespearian octave linked to Petrarchan sestet. Weeks lists seven examples of the first scheme, none of the second, one of the third. Hardy: 'Revulsion', i. 17, 'She, to Him I', i. 18, 'She, to Him IV', i. 20, 'To an Actress', i. 286, all of these in the first scheme; 'Hap', i. 10, 'She, to Him II', i. 19, 'Her Definition', i. 269, 'The Minute before Meeting', i. 287, 'Discouragement', iii.

155, all in the second scheme; ' "In vision I roamed" ', i. 10 in the third scheme. All of these sonnets were written in 1866, except 'Discouragement' (d. 1865–7), 'To an Actress' (d. 1867), and 'The Minute Before Meeting' (d. 1871).

abab cdcd efef gg Shakespearian sonnet. Palgrave, 3 ('When I have Seen by Time's Fell Hand Defaced'), 4 ('Since Brass, nor Stone, nor Earth, nor Boundless Seas'), 10 ('Being Your Slave, What Should I Do But Tend'), 11 ('How Like a Winter Hath My Absence Been'), 14 ('To Me, fair Friend, You Never Can Be Old'), 31 ('Farewell! Thou Art Too Dear for My Possessing'), 48 ('If Thou Survive My Well-contented Day'), 49 ('No Longer Mourn for Me When I am Dead'), 60 ('Tired with All These, for Restful Death I cry'), are the sonnets Hardy marks in his copy.

Hardy's 'Her Confession', i. 285; also 'She, to Him III', i. 19, 'From Her in the Country', i. 284, 'To an Impersonator of Rosalind', i. 286, all dated between 1865 and 1867.

abab cddc bce bce Fairly unusual octave, with a Petrarchan sestet.

Hardy's 'A Confession to a Friend in Trouble', i. 12, dated 1866. This combination of octave and sestet occurs only twice before, according to Weeks, and perhaps never before in this interlocking way.

abba abba cdcdcd Petrarchan sonnet; the last two examples
    cde cde have Wordsworthian sestets (see Schipper,
    cdedce ii. 872); the others have common
    cddccd Petrarchan sestets. For the Petrarchan
    cdc ddc sestets, see Palgrave, 70 (Milton's 'When the Assault Was Intended to the City'),

166 (Keats's 'On First Looking into Chapman's Homer'), 266 (Wordsworth's 'To Sleep'), all in the first scheme; 71 (Milton's 'On His Blindness'), 77 (Milton's 'To Cyriack Skinner'), both in the second scheme; 214 (Wordsworth's 'When I have Borne in Memory') in the third scheme.

Hardy's 'Zermatt: To the Matterhorn' (d. 1897), i. 138 in the first scheme; 'At a Lunar Eclipse', i. 149 (on dating, see Chapter 2, n. 2 above), ' "We are getting to the end" ' (published 1928), iii. 273, 'Thoughts from Sophocles' (d. 1895), iii. 307, all in the second scheme; ' "Often when warring" ' (d. 1915), ii. 298 in the third scheme; 'Rome: Building a New Street in the Ancient Quarter' (d. 1901), i. 135 in the fourth scheme; 'The Pity of It' (d. 1915), ii. 294 in the fifth scheme.

abba abba cdcdee   Petrarchan form, like Cowper's 'To Mary Unwin' (Palgrave, 161), Arnold's 'To a Republican Friend', or several examples in Sidney's *Astrophel and Stella*.
Hardy's 'By the Barrows' (p. 1901), i. 317.

ˆabba abba cdceed   Fairly unusual Petrarchan form, like Milton's 'To Mr. Lawrence' (Palgrave, 76).
Hardy's 'The Sleep-Worker' (p. 1901), i. 156.

abba abba cddc ee   Petrarchan form, like Wordsworth's 'Admonition to a Traveller' (Palgrave, 248).
Hardy's 'Rome: On the Palatine' (p. 1901), 68, 'A Church Romance' (p. 1906), i. 306.

abba acca adad ee   This combination of Wordsworthian octave and Shakespearian couplet is fairly common, though not perhaps in this interlocking way.

Hardy's 'Barthélémon at Vauxhall', ii. 331, dated 1921.

abba acca dedeed dedeff deeded deed ff def def defdfe	'Wordsworthian' sonnet modelled after Petrarch; all the schemes are fairly common. The last example is the scheme of 'It is a Beauteous Evening' (Palgrave 261). Hardy's 'At a Bridal' (d. 1866), i. 11, 'Departure' (d. 1899), i. 116 in the first scheme; 'The Schreckhorn' (perhaps begun 1897, d. 1906), ii. 30 in the second scheme; 'In the Old Theatre, Fiesole' (d. 1901), i. 134 in the third scheme; 'Embarcation' (d. 1899), i. 116 in the fourth scheme; 'On the Belgian Expatriation' (d. 1914), ii. 292, 'In Time of Wars and Tumults' (d. 1915), ii. 294, 'A Call to National Service' (d. 1917), ii. 300, all in the fifth scheme; 'To a Lady' (d. 189–), i. 85 in the sixth scheme.
abba cbbc deedde	This is Hardy's most unusual form. The combination of octave and sestet, and their interlocking rhymes, seems to be unique to him. Hardy wrote in 1918: 'I do not think that a departure from the customary Italian form is at all undesirable, though that form goes on year after year among so many poets' (*Collected Letters*, v. 276). Hardy's 'A Wet Night', i. 332, published 1909.

The following poems from the series, 'Satires of Circumstance', are strange amalgamations: fourteen-line tetrameter poems, the first two in what Trevor Johnson (at the 1986 Thomas Hardy conference) has called 'inverted sonnet' rhyme schemes; the last is in a somewhat more conventional rhyme scheme:

ababcc dededeff	Hardy's 'Over the Coffin', ii. 148.
ababcc dedeffgg	Hardy's 'At the Altar-Rail', ii. 145.
ababccdd efefgg	'Hardy's 'In the Cemetery', ii. 143.

'Blank Verse Sonnet'   Wordsworth referred to a 'perfect sonnet
                              without rhyme' in *Paradise Lost*, 'essentially
                              a sonnet in unity of thought' (the last
                              phrase apparently added by Crabb
                              Robinson to Wordsworth's account). Lee
                              Johnson building on this passage has
                              traced several imbedded blank verse sonnets
                              in Milton and Wordsworth's blank verse
                              epics (see *Wordsworth's Metaphysical Verse*
                              (Toronto, 1982)). There are about 10
                              imbedded 14 line units in the blank verse of
                              the Dynasts, 2 of which occur early and late
                              in the work and are potentially blank verse
                              sonnets in syntax, punctuation, and sense:
                              'These are the Prime Volitions'. p. 7, which
                              follows a rhyme scheme suggesting the
                              Petrarchan division of 8 and 6 lines: *abca
                              defg hhiijj*: also see 'Why did the death-
                              drop', pp. 519–20.

CLASSICAL EXPERIMENTS

In the context of heightened interest and speculation concerning
classical metres, Hardy more or less began his poetic career by
imitating classical verse. The first poem of *Wessex Poems* is 'The
Temporary the All', later subtitled '(Sapphics)' in Hardy's *Selected
Poems* of 1916. It is another example of how powerfully Hardy was
affected by Swinburne's *Poems and Ballads* with its poem entitled
'Sapphics'. Hardy marked this poem in his copy of Swinburne, and
scanned two of its lines:

> Āll thĕ nīght slĕep | cāme ˙ nŏt ŭp | ŏn mў ēyelĭds
> Sāw thĕ whīte īmplācăblĕ Āphrŏdītĕ

In his *Philological Grammar* (276) William Barnes had quoted Isaac
Watts's famous imitation of sapphics. The sapphic stanza imitated in
'The Temporary the All' is a four-line stanza in falling rhythm,
consisting of 3 five-foot hendecasyllabic lines and a final two-foot line
called an 'adonic'. The form is most recognizable by the dactyl in the
third foot of the long lines and in the first foot of the short lines. Fig. 4
contains a 'verse skeleton', like those in Noël's *Gradus*, of one common
form of sapphics.

	1	2	3	4	5
First 3 lines	– ◡	– – – ◡	– ◡ ◡	– ◡	– ◡ – –
Fourth line	– ◡ ◡	– –			

Fig. 4

(Hardy's scansion also indicates a caesura point.) Thus:

> Change and chancefulness in my flowering youthtime
> Set me sun by sun near to one unchosen;
> Wrought us fellowlike, and despite divergence,
>   Fused us in friendship.

Hardy continued to experiment with classical metres throughout his life. 'The Collector Cleans His Picture' (ii. 388) has several lines which are like sapphic lines. Ford Madox Ford (448) claimed that in a conversation Hardy pointed to 'Sapphics' in *The Dynasts*. The poem in question was probably the 'Chorus of Rumors' (243), which went through an extraordinary number of revisions from one printed version to the next, showing Hardy trying to get it right. The revisions show Hardy repeatedly attacking the problem of how to make the accentual pattern match the Latin quantitative norm. In the British Museum manuscript, in the first edition and in the editions of 1906 and 1909, the first stanza begins:

> Tén of thĕ níght ĭs Tálăvĕră tóllĭng:
> Nów dŏ Rúffin's ránks cŏme súrgĭng úpwărd.

The first line lacks the required third foot dactyl and begins with a dactyl instead of a trochee. The second line also lacks the third foot dactyl and the required eleven syllables. The 1910 edition shows Hardy correcting these violations of the pattern:

> Tálăvĕră tóngues ĭt ăs tén ăt níght-tĭme:
> Nów cŏme Rúffin's sláughtĕrĕrs súrgĭng úpwărd.

The next revision, for the 1913 Wessex edition and the 1920 Anniversary edition of Hardy's works, rewrites the first line and keeps the dactyl right.

Knélls ŏf níght ĭs véxt Tălăvĕră tóngŭing:
Now come Ruffin's slaughterers surging upward.

Perhaps Hardy felt that 'night is' is closer to the desired spondee.
Also, the change in positioning of 'Talavera' shows how the English
line pushes around its stresses. The Mellstock edition of 1920 and the
first thin paper edition of 1923 show yet another change:

Tálăvĕră tóngues ĭt ăs tén o'ˇthe níght-tĭme:
Now come Ruffin's slaughterers surging upward.

We have come full circle, back to ten o'clock, but now Hardy must
squeeze 'Ten of the', a dactyl in the first verse, into a trochee here by
means of a traditional contraction; 'Ten o'the'. He also gives up on
the spondee in the second foot. Such are the writhings of the accentual
imitator of Latin verse.

This is all good fun; but there is significance in the fact that Hardy
fiddled with sapphic stanza throughout his career. He told Elliot
Felkin that 'he often wrote verse in sapphics but intentionally not
quite correct'. This reflects something about the history of the sapphic
form. The development of the form through Horace up through the
centuries of Latin hymnody and finally into English and German
hymnody, is an interesting one. While the form is adapted in various
ways, the quantitative rhythms gradually change to accentual ones,
and rhyme is increasingly used.

But the influence of the sapphic stanza was not confined to hymns,
or even to the use of the precise stanza form. Concluding a stanza with
a short two-beat line tended to be considered a sapphic effect. Thus
Saintsbury, commenting on the Elizabethan miscellany *The Phoenix
Nest*, cited the 'Adonic of a Sapphic' in the short line of:

All day I weep my weary woes,
    That when that night approacheth near,
And every one his eyes doth close,
    And passed pains no more appear—
        I change my cheer.   (ii. 132)

This particular form is used in *Hymns Ancient and Modern*, 629, which
Hardy marked and listed in the back of his copy. It is also used in
Swinburne's 'Félise' and finally, though with more trisyllabic substitutions, in Hardy's 'At Castle Boterel' (ii. 63) and 'Old Furniture' (ii.
227). Saintsbury also noted the "Sapphics, at least in intention' of

The fatal star, that at my birthday shined,
Were it of Jove, or Venus in her brightness,
All sad effects, sour fruits of Love, divined
    In my love's lightness.   (ii. 132)

Examples of this form can be found in many of the poetry anthologies Hardy owned and it was used by him in 'After the Last Breath' (i. 326).

Thus while sapphics in their strict form were imitated in English only on occasion, the quatrains which loosely derived from them were widely practised. In calling many of his poems 'Sapphics but intentionally not quite correct', Hardy is evoking an enormous historical prospect. He is also calling our attention to the point where his classical experiments and his most frequently used verse form, the common measure of hymn and ballad, come out of a common tradition.

The chronology and placement of his early poems show Hardy's development of these roughened sapphics quite clearly. After 'The Temporary the All' '(Sapphics)' in *Wessex Poems* appears 'Amabel', dated 1865, an English development of the sapphic idea into a rhymed iambic quatrain with its final two-beat refrain line: $a^3a^3b^3b^2$ d.r. Four poems later is 'Postponement', dated 1866, which is more faithful to the classical model because of its largely dactylic rhythm and its 'adonic' fourth line: $a^4a^4a^4B^2$ t.f. (i.e. a triple falling rhythm with the fourth line a refrain line). In the same year Hardy wrote 'Dream of the City Shopwoman' (ii. 379), another rhymed iambic quatrain with a two-beat fourth line. The form imitates that of Wyatt's 'Forget Not Yet', a Palgrave selection (21): $a^4a^4a^4b^2$ d.r. He also wrote in this year 'The Musing Maiden' (iii. 244), an iambic quatrain of three tetrameter lines concluding with a trimeter line. These last two poems are less interesting technical variations on the sapphic form; for this reason perhaps, Hardy reserved them for later publication. The next example, however, is very interesting. 'Neutral Tones', dated 1867, comes two poems after 'Postponement' in *Wessex Poems*. It is also a quatrain of three tetrameter lines with a final trimeter line. But it has a much more interesting mixture of iambs and dactyls, perhaps reminiscent to Hardy of the dactyls and trochees in sapphics. E. B. Browning had used a similar form in 'The Deserted Garden'. About this time Hardy wrote another poem with a shortened final line: the quatrain of 'Her Father' (i. 273) somewhat like 'The Musing Maiden' but in the form: $a^4b^4a^4b^3$ d.t.r. It was kept back for later publication. 'On the Departure Platform' (i. 271) may also be from this time; it imitates a common form found in such poems Hardy knew as Newman's 'O Thou, of Shepherds', Byron's 'On This Day I Complete', and Pope's 'Happy the Man, Whose Wish and Care' (Palgrave, 118): $a^4b^4a^4b^2$ d.r., though Hardy uses more trisyllabic variation.

While Hardy experiments with the sapphic stanza in its strict and loose forms, he also makes other classical experiments. In the 1870s, in 'After Reading Psalms xxxix, xl, etc.' (ii. 484), he imitates middle English Latin-tag verse which mingles English lines and Latin tags. The Latin line is put into an accentual rhythm:

> Simple was I and was young;
> Kept no gallant tryst, I;
> Even from good words held my tongue,
> *Quoniam Tu fecisti!*

The form of the poem is a close likeness to the first four lines of 'Of on that is so fayr and bright' in the Egerton manuscript. The Latin tag line may show Hardy's awareness of the rich heritage of his quatrains. In the rhyme scheme and number of beats, we may see the emergence of 'common measure', here somewhat hidden behind the falling rhythm associated with sapphic imitations.

When Hardy returned to a full-time career as poet in the 1890s, he began with a reconsideration of classical metres and their relation to English verse. His 'In Tenebris iii' (i. 208, d. 1896) might be considered a loose accentual-syllabic imitation of the dactylic hexameter or 'heroic' line. I cannot find the 'Heroics' Ford Madox Ford (448) claimed Hardy pointed to in *The Dynasts*. This line has been the most imitated of classical metres in English. In his 1896 edition of Donne's poems, Hardy attempts to divide, in the following manner, a line from Horace (Satires i. 2, l. 119), which Saintsbury's introduction characterized as a 'hideous hexameter' (p. xx):

> Non ego |, namquḗ para | bilem a | mo Vene | rem faci | lemque

It consists of six feet in falling rhythm, most recognizable by the dactyl in the penultimate foot and the concluding trochee or spondee (Fig. 5).

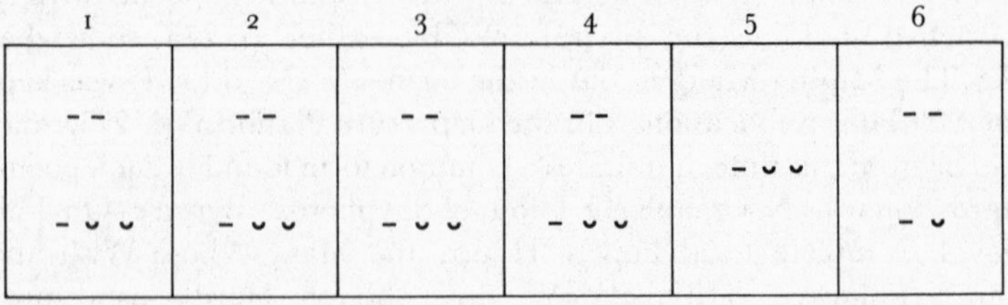

FIG. 5

The history of those imitations, from the sixteenth-century Areopagus poets through Clough and Longfellow, illustrates how English imitation of quantities became simply English substitution of accentual-syllabic units for feet. Hardy's 'In Tenebris III' (i. 208) might be called an accentual-syllabic imitation, but in a very loose sense: in the first stanza, for example, the sixth foot in the first and last lines is monosyllabic, the penultimate dactyl is missing in the third line and the lines are rhymed.

About the time he wrote this poem, Hardy may have been puzzled by the anachronism of these imitations. He may have read William Stone's *On the Use of Classical Metres in English* (1899) at this time, with its controversial argument on the close relation between Greek and English quantities. The tide of classical experimentation had lapsed in the 1870s (6 items in Omond's list) and 1880s (10 items), but then had swelled once more with 20 items in the 1890s (including Hardy's 'Sapphics') and another 20 in the 1900s, including many experiments by Bridges and reviews of these. 'After the battles of the sixties there was some rest over the prosodic world . . . till at last a youthful champion [Stone] . . . waked the lists again' (Saintsbury, iii. 429). The fact that Stone and this new classical awakening coincided with a renewed interest in classical experimentation may have been part of the reason for an interesting incident recorded in the *Life* for 1900. Hardy describes the incident shortly after giving us his *Later Years* assessment of his metres. He has just cited the 'confusion of thought to be observed in Wordsworth's teaching'. He continues:

For some reason he spent time while here in hunting up Latin hymns at the British Museum, and copies that he made of several have been found, of dates ranging from the thirteenth to the seventeenth century, by Thomas of Celano, Adam of S. Victor, John Mombaer, Jacob Balde, etc. That English prosody might be enriched by adapting some of the verse-forms of these is not unlikely to have been his view. (*Life*, 306.)

I suspect this is one example (his visit to the Gloucester transept is another—see Taylor, *Hardy's Poetry*, 53) of Hardy researching the historical sources of his art. Bridges, we recall, says that many late Latin hymns kept only the accentual speech rhythms of the older quantitative hymnody 'without attention to the prosody that originally provoked and sustained them, and their poems give the flesh without any skeleton' (*Milton's Prosody*, 1921 edition, 86). By the thirteenth century, the Latin accentual hymn was well established, and was one of the sources of English ballad and hymn rhythms. In

other words, by consulting its sources, Hardy may have increasingly
realized the true nature of his own prosody, the relation of
'accentuation' to 'long and short syllables' (see above, Chapter 2).
The copies Hardy made in the British Museum were inserted in his
1871 copy of *Sequences from the Sarum Missal*, translated by C. B.
Pearson. Hardy imitated two of the forms in his poems 'Sine Prole'
(iii. 30), subtitled 'Mediaeval Latin Sequence-Metre', and 'Genitrix
Laesa' (iii. 88). The latter is an imitation of a form used in Adam of St
Victor's 'Officium Beatae Mariae' and in other examples from the
Sarum Missal. Thus Hardy:

> Nature, through these generations
> You have nursed us with a patience
> Cruelly crossed by malversations,
>   Marring mother-ministry
> To your multitudes, so blended
> By your processes, long-tended,
> And the painstaking expended
>   On their chording tunefully.

Hardy is here making an English accentual-syllabic imitation of a
Latin accentual-syllabic rhythm as it emerged from the old quantitative rhythms and influenced and was influenced by the new
'irrational' rhythm.

According to F. M. Ford's somewhat unreliable recollections (448),
Hardy used sapphics and alcaics in *The Dynasts*. We have seen the
sapphics. The alcaic stanza is a quatrain in falling rhythm, perceived
in English accentual imitations as a complex four-foot metre, perhaps
most recognizable by the double dactyls concluding the first two lines,
and beginning the last line. A common form is that in Fig. 6.

	1	2	3	4	5
First 2 lines	–	– ᵕ	– –	– ᵕ ᵕ	– ᵕ ᵕ
Third line	–	– ᵕ	– –	– ᵕ	– –
Fourth line	– ᵕ ᵕ	– ᵕ ᵕ	– ᵕ	– ᵕ	

Fig. 6

The form was imitated by Milton and Tennyson (in 'Milton'), both
with unusual regard for quantities. In his *Philological Grammar* (276),
Barnes quotes an example by Sidney. The form may be very loosely

imitated in Hardy's 'Feeble-framed dull unresolve' (*Dynasts*, 16), the closest approximation of 'alcaics' Ford said Hardy noted in the work. The imitation, if indeed Hardy intended such, is very loose: it at least keeps the double dactyl effect. Indeed (or nevertheless) all the lines fall into a dactylic tetrameter pattern. There are striking revisions in the printed versions which may indicate Hardy trying to hone the imitation (as is the case with the 'Chorus of Rumors' discussed above). The revision of the third line, for example, from

> Whence an untactical torpid despondency

to

> Whence the grey glooms of a ghost-eyed despondency

(in the 1904 London reprint of the 1903 edition) seems to slow down the line, but the result is still dubiously 'alcaic'.

Like the sapphic stanza, the alcaic stanza has a long complex history from fragments of Alcaeus through Horace into Latin hymnody from whence it diffuses into the broad stream of the English quatrain. Culler, *The Poetry of Tennyson* (New Haven, 1977), 81, points out that Tennyson thought of his *In Memoriam* stanza as emulating Alcaeus. So Hardy may have associated alcaics with other of his quatrains and construed them as 'roughened alcaics'. Indeed roughened alcaics and roughened sapphics may have been roughly equivalent in his mind and in the mind of many late versifiers.

Hardy's last clearly recognizable classical imitation occurs in the poem 'Aristodemus the Messenian' (iii. 181) subtitled 'Dramatic Hendecasyllabics'. Published in 1928 in his last volume of poetry, it is the strictest of Hardy's classical experimentations since the first poem of his first volume, *Wessex Poems*. The line is eleven syllables, five feet, in falling rhythm, most easily recognizable by the dactyl which always occurs in the second foot. A common scheme is that in Fig. 7.

Fig. 7

These two imitations, 'The Temporary the All' and 'Aristodemus the Messenian', bracket Hardy's metrical career. Like the first one, the last one shows the influence of Swinburne—Swinburne's 'Hendeca-syllabics', 'In the Month of the Long Decline of Roses', which Hardy marked in his copy of *Poems and Ballads*, First Series.

Ezra Pound made a perceptive observation about Hardy in the appendix to his anthology, *Confucius to Cummings*: 'Every conventional fashion, all the undergraduate efforts to use ancient metres—no man ever had so much Latin and so eschewed the least appearance of being a classicist' (325).

Bibliography

ABERCROMBIE, LASCELLES, *Principles of English Prosody* (London, 1923).

ABBOTT, EDWIN A., *A Shakespearean Grammar* (London, 1869).

—— and J. R. SEELEY, *English Lessons for English People* (Boston, 1872).

ABRAMS, MEYER H., 'The Correspondent Breeze: A Romantic Metaphor', in id. (ed.), *English Romantic Poets* (New York, 1960).

—— 'Structure and Style in the Greater Romantic Lyric', in Hilles and Bloom (eds.), *From Sensibility to Romanticism* (Oxford, 1965).

ALDEN, RAYMOND, *English Verse* (New York, 1903).

—— 'The Mental Side of Metrical Form', *MLR* 9 (1914), 298–308.

ARCHER, RICHARD, *Secondary Education in the Nineteenth Century* (London, 1966).

ARNOLD, MATTHEW, *On Translating Homer* (1861); repr. in R. H. Super (ed.), *On the Classical Tradition* (Ann Arbor, 1960).

ATTRIDGE, DEREK, *The Rhythms of English Poetry* (London, 1982).

—— *Well-weighed Syllables: Elizabethan Verse in Classical Metres* (Cambridge, 1974).

AUBIN, ROBERT, *Topographical Poetry in XVIII-century England* (New York, 1936).

AUDEN, W. H., 'A Literary Transference', *Southern Review* 6 (1940–1), 78–86.

—— *Making, Knowing, and Judging* (1956); repr. in *The Dyer's Hand and Other Essays* (New York, 1968).

BAILEY, JAMES O., *The Poetry of Thomas Hardy* (Chapel Hill, 1970).

Ballad Minstrelsy of Scotland, 2nd edn. (Glasgow, 187–). Original edn. pub. 1871 (Hardy's copy, Dorset County Museum).

BARHAM, THOMAS, 'On Metrical Time, or, the Rhythm of Verse, ancient and modern', *Transactions of the Philological Society* (1860–1), 45–62.

BARNES, WILLIAM, *A Philological Grammar* (London, 1854).

—— *Select Poems*, ed. Thomas Hardy (London, 1908).

BATE, WALTER JACKSON, *The Stylistic Development of Keats* (New York, 1945).

BAYFIELD, M. A., 'Our Traditional Prosody and an Alternative', *MLR* 13 (1918), 157–82.

—— 'Shakespeare's Versification and the Early Texts', *TLS* (23 May 1918), 242; (6 June), 265; (13 June), 277. See also discussion following in *TLS* (27 June), 301; (4 July), 313; (11 July), 325; (1 Aug.), 361–2; (5 Sept.), 417; (26 Sept.), 455–6.

BAYLY, THOMAS HAYNES, *Songs, Ballads, and Other Poems*, 2 vols. (London, 1844) (Hardy's copy, Wreden catalogue).

BEDE, 'De Rithmo' (*c*.691–703). Part 23 of *De Arte Metrica*, in *Bedae Venerabilis Opera*, Part I: *Opera Didascalia* (Turnholti 1975), 138–9.

The Best of Friends: Further Letters to Sydney Carlyle Cockerell, ed. V. Meynell (London, 1956).

BLUNDEN, EDMUND, *Thomas Hardy* (London, 1942).

BRIDGES, ROBERT, 'A Letter to a Musician on English Prosody', *Musical Antiquary*, 1 (1909), 15–29; rev. and repr. in *Collected Essays*, No. 15 (London, 1933).

—— 'Humdrum & Harum-Scarum: A Lecture on Free Verse' (*London Mercury*, 1922), repr. in *Collected Essays*, No. 2 (London, 1928).

—— *Milton's Prosody*, Rev. final edn. (Oxford, 1921). Earlier edns. 1893, 1901. Rev. form of the two following essays:

—— *On the Prosody of Paradise Regained and Samson Agonistes* (Oxford, 1889).

—— 'On the Elements of Milton's Blank Verse in Paradise Lost', in H. C. Beeching (ed.), *Paradise Lost: Book I* (Oxford, 1887).

—— *Selected Letters*, ed. D. Stanford, 2 vols. (Newark, 1983).

—— 'Wordsworth and Kipling' (originally in *TLS*, 29 Feb. 1912); repr. in *Collected Essays*, No. 13 (London, 1933).

BRINK, BERNARD TEN, *Chaucers Sprache and Verskunst* (Leipzig, 1884), 2nd edn. 1899, trans. M. B. Smith as *The Language and Meter of Chaucer* (London, 1901).

BROGAN, T. V. F., *English Versification, 1570–1980: A Reference Guide with a Global Appendix* (Baltimore, 1981).

—— *Three Models for English Verse*, unpub. manuscript.

BROOKS, CLEANTH, 'The Language of Poetry, Some Problem Cases'. *Archiv für das Studium Der Neuren Sprache und Literaturen*, 203 (1966–7), 401–14.

—— 'A Note on Thomas Hardy', *Hopkins Review*, 5 (1952), 68–79.

BROWNE, W. H., 'Certain Considerations touching the Structure of English Verse', *MLN* 4 (1889), 193–202.

BYSSHE, EDWARD, *The Art of English Poetry*, introd. Dwight Culler (Los Angeles, 1953; originally pub. London, 1702).

CAYLEY, CHARLES B., 'The Pedigree of English Heroic Verse', *Transactions of the Philological Society* (1867), 43–54.

—— 'Remarks and Experiments on English Hexameters', *Transactions of the Philological Society* (1862–3), 67–85.

CHILD, FRANCIS J., *English and Scottish Ballads*, 8 vols. (Boston, 1857–9).

—— *The English and Scottish Popular Ballads*, 10 parts (London, 1882–98).

—— 'Observations on the Language of Chaucer', *Memoirs of the American Academy of Arts and Sciences*, NS 8 (1863), 445–99. Repub. with some modification in Ellis, *On Early English Pronunciation*, i (1869), q.v.

COHEN, HELEN, *Lyric Forms from France* (New York, 1922).

COLERIDGE, MARY, *Poems*, ed. Henry Newbolt (London, 1910) (Hardy's copy, Dorset County Museum).

COLERIDGE, SAMUEL TAYLOR, *Biographia Literaria*, ed. J. Engell and W. J. Bate (Princeton, 1983).

—— *Miscellaneous Criticism*, ed. T. M. Raysor (Cambridge, 1936).

COX, J. STEVENS (ed.), *The Library of Thomas Hardy*: repr. of the Hodgson 1938 sales catalogue, in *Monographs on the Life, Times and Works of Thomas Hardy*, No. 52 (Guernsey, 1969).

COXON, PETER, 'Thomas Hardy: "The Voice" and Horace: Odes II, XIV', *Thomas Hardy Society Review*, 1 (1983), 291–3.

CULLER, A. DWIGHT, 'Edward Bysshe and the Poet's Handbook', *PMLA* 63 (1948), 858–85.

DARLEY, GEORGE, *Complete Poetical Works*, ed. Ramsay Coles (London, 1908) (Hardy's copy, Univ. of Texas).

DAVIE, DONALD, *Thomas Hardy and British Poetry* (New York, 1972).

DE LA MARE, WALTER, *Poems* (London, 1906) (Hardy's copy, Dorset County Museum).

DE SELINCOURT, BASIL, 'English Prosody', *Quarterly Review*, 215 (July, 1911), 68–96.

—— chapters 14–17 in M. Sturge Henderson (Mary Gratton), *George Meredith* (New York, 1907).

DIBDIN, T. (ed.), *Songs of the Late Charles Dibdin*, 3rd edn. (London, 1872) (Hardy's copy, Dorset County Museum).

DONNE, JOHN, *Poems*, ed. E. K. Chambers, introd. G. Saintsbury, 2 vols. (London, 1896) (Hardy's copy, Dorset County Museum).

The Dunbar Anthology 1401–1508 A.D., ed. Edward Arber (London, 1901) (Hardy's copy, Dorset County Museum).

ELIOT, GEORGE, *A Writer's Notebook 1854–1879*, ed. Joseph Wiesenfarth (Charlottesville, Va., 1981; first printed in 1980).

ELLIS, ALEXANDER, *On Early English Pronunciation*, 5 vols. (London, 1869–89).

—— *Practical Hints on the Quantitative Pronunciation of Latin* (London, 1874).

ERHARDT-SIEBOLD, ERIKA VON, 'Some Inventions of the Pre-Romantic Period', *Englische Studien*, 66 (1932), 347–63.

FELKIN, ELLIOTT, 'Days with Thomas Hardy', *Encounter*, 18 (Apr. 1962), 27–33.

FIROR, RUTH, *Folkways in Thomas Hardy*, (Philadelphia, 1931).

FLETCHER, ROBERT, 'The Metrical Forms Used by Certain Victorian Poets', *JEGP* 7 (1908), 87–91.

FOGG, PETER, *Elementa Anglicana; or, the Principles of English Grammar*, 2 vols. (Stockport, 1792–6).

FORD, FORD MADOX, 'Thomas Hardy', *American Mercury*, 38 (1936), 438–48.

FORMAN, H. BUXTON (ed.), *Note Books of Percy Bysshe Shelley* (Boston, 1911).

Friends of a Lifetime: Letters to Sydney Carlyle Cockerell, ed. V. Meynell (London, 1940).

FROST, MAURICE (ed.), *Historical Companion to Hymns Ancient and Modern* (London, 1962).

FURNIVALL, F. J., *The Succession of Shakespeare's Works and The Use of the Metrical Tests in Settling It* (London, 1874).

FUSSELL, PAUL, *Poetic Meter and Poetic Form*, rev. edn. (New York, 1979; first edn., 1965).

—— *Theory of Prosody in Eighteenth-century England* (New London, Conn., 1954).

GASCOIGNE, GEORGE, 'Certayne Notes of Instruction concerning the making of verse or ryme in English' (1575), in G. Gregory Smith (ed.), *Elizabethan Critical Essays*, vol. i (London, 1904).

GERBER, HELMUT, and EUGENE DAVIS, *Thomas Hardy: An Annotated Bibliography of Writings About Him*, 2 vols. (De Kalb, Ill., 1973, 1983).

GILDON, CHARLES, 'New Prosodia', in John Brightland, *A Grammar of the English Tongue* (London, 1711).

GITTINGS, ROBERT, *Young Thomas Hardy* (London, 1975).

—— *The Older Hardy* (London, 1978).

GOING, WILLIAM, *Scanty Plot of Ground: Studies in the Victorian Sonnet* (The Hague, 1976).

GOODELL, THOMAS, *Chapters on Greek Metric* (New Haven, 1901).

GORDON, ALFRED, 'The New Prosody: a Rejoinder', *Canadian Bookman* (April 1920), 33–45.

—— 'What is Poetry?—A Synthesis of Modern Criticism', *Canadian Bookman* (Apr. 1919), 39–46.

GOSSE, EDMUND, *Firdausi in Exile and Other Poems* (London, 1885) (Hardy's copy, Cox catalogue).

—— *Life and Letters*, ed. Evan Charteris (New York, 1931).

—— *On Viol and Flute: Selected Poems* (London, 1890) (Hardy's copy, Cox catalogue).

—— 'A Plea for Certain Exotic Forms of Verse', *Cornhill Magazine* 36 (1877), 53–71.

Gradus ad Parnassum, ed. Francis Noël, new edn. (Paris, 1882).

GRIGSON, GEOFFREY, *The Harp of Aeolus* (London, 1948 (1947)).

GROSS, HARVEY, *Sound and Form in Modern Poetry* (Ann Arbor, Mich., 1964).

GUEST, EDWIN, *A History of English Rhythms* 2 vols. (London, 1838).

—— *A History of English Rhythms* (1838), new edn. ed. Walter W. Skeat (London, 1882) (Hardy's copy, Cox catalogue).

GUMMERE, FRANCIS, *The Beginnings of Poetry* (New York, 1901) (Hardy's copy, Mathews catalogue).

—— *A Handbook of Poetics* (Boston, 1885).

—— 'The Translation of *Beowulf*, and the Relation of Ancient and Modern English Verse', *American Journal of Philology*, 7 (1886), 46–78.

GUNN, THOM, 'Hardy and the Ballads', *Agenda*, 10 (1972).

HALLAM, HENRY, *Introduction to the Literature of Europe in the Fifteenth, Sixteenth, and Seventeenth Centuries*, 4 vols. (London, 1837–9).

HALLE, MORRIS, and SAMUEL J. KEYSER, 'Chaucer and the Study of Prosody', *College English*, 28 (1966), 187–219.

—— *English Stress* (New York, 1971).

HARDY, EVELYN, *The Countryman's Ear, and Other Essays on Thomas Hardy* (Padstow, 1982).

—— *Thomas Hardy* (New York, 1954).

HARDY, THOMAS, *The Collected Letters of Thomas Hardy*, ed. Richard Purdy and Michael Millgate, vol. i (1840–1892), ii (1893–1901), iii (1902–1908), iv (1909–13), v (1914–19), vi (1920–5) (Oxford, 1978, 1980, 1982, 1984, 1985, 1987).

—— *Complete Poems*, variorum edn. ed. James Gibson (London, 1979).

—— *The Complete Poetical Works of Thomas Hardy*, ed. Samuel Hynes, 3 vols. (Oxford, 1982, 1984, 1985).

—— *The Dynasts* (London, 1965). Originally pub. in 3 vols. (London, 1903 (1904), 1905 (1906), 1908).

—— *The Early Life of Thomas Hardy 1840–1891 By Florence Emily Hardy* (London, 1928).

—— *The Famous Tragedy of the Queen of Cornwall* (London, 1923).

—— *The Later Years of Thomas Hardy 1892–1928 by Florence Emily Hardy* (London, 1930).

—— *The Life and Work of Thomas Hardy*, ed. Michael Millgate (Athens, Ga., 1985).

—— *The Life of Thomas Hardy 1840–1928 by Florence Emily Hardy*, combined edn. (London, 1962).

—— *The Literary Notebooks of Thomas Hardy*, ed. Lennart Björk, 2 vols. (New York, 1985): includes *Literary Notes* I, II, III, and the so-called ' "1867" Notebook', all originally pub. in *Original Manuscripts*.

—— *Memoranda Book I and II*: pub. in *Original Manuscripts*, Reel 9 and in *Personal Notebooks*, ed. Richard Taylor.

—— *The Original Manuscripts and Papers* (Wakefield, EP Microform Ltd., 1975):

    Reel 5: *Poetry, Essays and Short Stories*;

    Reel 6: *Scrapbooks* ('Personal Reviews', etc.);

    Reels 7 and 8: *Drafts for Biography*;

    Reel 9: *Memoranda, Diaries, and Notebooks*. (Also includes the unpublished 'Facts' notebook, entitled 'Commonplace Books III').

    Reel 10: *Music Books, Paintings, and Drawings*.

—— *Personal Notebooks*, ed. Richard H. Taylor (London, 1979).

—— *Personal Writings, Thomas Hardy's*, ed. Harold Orel (Lawrence, 1966).

—— *Real Conversations*, by William Archer (London, 1904).

—— *Selected Poems* (London, 1916).

—— *Talks with Thomas Hardy at Max Gate, 1920–1922*, by Vere Collins (London, 1928).

—— *The Wessex Novels*, in *The Works of Thomas Hardy in Prose and Verse*, Wessex edn. (London, 1912–31). Includes the following novels cited in the text:

    Desperate Remedies, originally pub. 1871.

    Under the Greenwood Tree, 1872.

    A Pair of Blue Eyes, 1873.

    Far from the Madding Crowd, 1874.

    The Hand of Ethelberta, 1876.

The Return of the Native, 1878.

The Mayor of Casterbridge, 1886.

The Woodlanders, 1887.

Tess of the d'Urbervilles, 1891.

Jude the Obscure, 1896.

The Well-Beloved, 1897.

HARTMAN, GEOFFREY, 'Wordsworth, Inscriptions, and Romantic Nature Poetry', in Hilles and Bloom (ed.) *From Sensibility to Romanticism* (Oxford, 1965).

HEDGCOCK, FRANK, 'Reminiscences of Thomas Hardy', *National and English Review*, 137 (1951), 220–8, 289–94.

HEGEL, GEORGE WILHELM FRIEDRICH, *Aesthetics: Lectures on Fine Art*, trans. T. M. Knox., 2 vols. (Oxford, 1975). Originally pub. in German in 1835–8; first trans. (in part) into French by M. Ch. Bénard as *Cours d' Esthétique*, 5 vols. (Paris, 1840–52); first trans. in part into English in 1867, *in toto* by F. P. B. Osmaston as *The Philosophy of Fine Art* (London, 1920).

—— *The Phenomenology of Mind*, trans. J. B. Baillie, 2nd edn. (London, 1931; first edn. 1910). Originally pub. in German in 1807; first trans. in part into English in 1868.

HEINE, HEINRICH, *Book of Songs*, trans. Charles G. Leland (New York, 1881) (Hardy's copy, Holmes Listing: Purdy Collection).

—— *Poems*, trans. Edgar Alfred Bowring (London, 1878) (Hardy's copy, Purdy, 117; Holmes Listing: Purdy Collection).

HENDRICKSON, G. L., 'Elizabethan Quantitative Hexameters', *Philological Quarterly*, 28 (1949), 237–60.

HENLEY, W. E. (ed.), *English Lyrics: Chaucer to Poe 1340–1809* (London, 1893) (Hardy's copy, Dorset County Museum).

HERBERT, GEORGE, *Poetical Works*, introd. G. Saintsbury, 2 vols. (London, 1893) (Hardy's copy, Dorset County Museum).

HICKSON, ELIZABETH, *The Versification of Thomas Hardy* (Philadelphia, 1931).

HOLLANDER, JOHN, *The Untuning of the Sky: Ideas of Music in English Poetry, 1500–1700* (Princeton, 1961).

—— *Vision and Resonance: Two Senses of Poetic Form* (New York, 1975).

—— 'Wordsworth and the Music of Sound', in G. Hartman (ed.), *New Perspectives on Coleridge and Wordsworth* (New York, 1972).

HOPKINS, GERARD MANLEY, *The Correspondence of Gerard Manley Hopkins and Richard Watson Dixon*, ed. Claude Abbott, rev. impression (London, 1955).

—— *Further Letters of Gerard Manley Hopkins, Including His Correspondence with Coventry Patmore*, ed. C. Abbott 2nd edn. (London, 1956).

—— H. House and G. Storey (eds.), *The Journals and Papers of Gerard Manley Hopkins* (New York, 1959).

—— *The Letters of Gerard Manley Hopkins to Robert Bridges*, ed. C. Abbott, rev. impression (London, 1955).

—— *Poems*, ed. W. H. Gardner and N. H. MacKenzie, 4th edn. (London, 1970).

HORACE, *Works*, trans. C. Smart, new edn. Theodore Buckley (London, 1859) (Hardy's copy, Colby College Library).

HOWE, IRVING, 'The Short Poems of Thomas Hardy', *Southern Review*, NS 2 (1966), 878–905.

HULLAH, JOHN (ed.), *The Song Book: Words and Tunes from the Best Poets and Musicians*, (London, 1866) (Hardy's copy, Dorset County Museum).

HULME, T. E., 'Lecture on Modern Poetry', repr. in M. Roberts, *T. E. Hulme* (Manchester, 1982).

HUNT, LEIGH, 'What is Poetry', in id., *Imagination and Fancy* (London, 1844).

HUTCHESON, FRANCIS, 'An Inquiry into the Original of Our Ideas of Beauty and Virtue' (1725), in Scott Elledge (ed.), *Eighteenth-Century Critical Essays*, vol. 1. (Ithaca, NY, 1961).

HYMES, DELL, 'Phonological Aspects of Style: Some English Sonnets', in T. Sebeok (ed.), *Style in Language* (Cambridge, 1960).

Hymns Ancient and Modern (1860–1).

—— with new appendix (London, 1868).

—— rev. and enlarged edn. (London, 1875).

—— with suppl. 1889 (Hardy's copy, Dorset County Museum).

—— new edn. (London, 1904).

—— old (1889) edn. with Metrical Index of Tunes (London, 1906).

HYNES, SAMUEL, *The Pattern of Hardy's Poetry* (Chapel Hill, 1961).

JAKOBSON, ROMAN, 'Closing Statement: Linguistics and Poetics', In T. Sebeok (ed.), *Style in Language* (Cambridge, 1960).

JEAFFRESON, JOHN C., *The Real Shelley*, 2 vols. (London, 1885) (Hardy's copy, Cox catalogue).

JESPERSEN, OTTO, *Linguistica: Selected Papers in English, French and German* (College Park, Md., 1970).

—— 'Notes on Metre' (1903 in Danish, 1933 in English trans.), in S. Chatman and S. Levin (eds.), *Essays on the Language of Literature* (Boston, 1967).

—— *Progress in Language* (London, 1894).

JOHNSON, H. A. T., 'Thomas Hardy and the Respectable Muse', *Thomas Hardy Yearbook*, No. 1 (1970), 27–42.

JOHNSON, SAMUEL, *Lives of the English Poets* (London, 1886; originally pub. 1781) (Hardy's copy, University of Texas).

—— 'Grammar of the English Tongue', *A Dictionary of the English Language*, 2 vols. (London, 1755).

JULIAN, JOHN (ed.), *A Dictionary of Hymnology*, rev. edn., 2 vols. (London, 1907).

KAMES, LORD HENRY HOME, 'Versification', in id., *Elements of Criticism*, vol. 2. (Edinburgh, 1762; Johnson Reprint, 1967).

KEATS, JOHN, *Poetical Works*, ed. W. M. Rossetti (London, 1872?) (Hardy's copy, Dorset County Museum).

KEBLE, JOHN, *The Christian Year* (London, 1860; originally pub. in 1827) (Hardy's copy, Dorset County Museum).

KEY, T. HEWITT, 'A Partial Attempt to Reconcile the Laws of Latin Rhythm with those of Modern Languages', *Transactions of the Philological Society* (1868–9), 311–51.

KING, ROBERT W., 'The Lyrical Poems of Thomas Hardy', *London Mercury* 15 (1926), 157–70.

KIPARSKY, PAUL, 'The Rhythmic Structure of English Verse', *Linguistic Inquiry*, 8 (1977), 189–247.

KITTREDGE, GEORGE LYMAN, *Observations on the Language of Chaucer's 'Troilus'*, Chaucer Society Publications, 2nd ser., No. 28 (London, 1891).

KNIGHT, W. F. JACKSON, *Accentual Symmetry in Vergil* (Oxford, 1950).

LANCELOT, CLAUDE, 'Breve Instruction sur les Regles de la Poësie Françoise', in id., *Quatre Traitez de Poësies Latine, Françoise, Italienne, et Espagnole* (Paris, 1663).

LANG, ANDREW, 'At the Sign of the Ship', *Longman's Magazine*, 36 (May 1900), 88–96.

LANIER, SIDNEY, *The Science of English Verse* (New York, 1880).

LARKIN, PHILIP, 'Philip Larkin Praises the Poetry of Thomas Hardy', *The Listener* (25 July 1968), 111.

LATHAM, ROBERT G., *The English Language*, 4th rev. edn. (London, 1855). Originally pub. in 1841.

LEAVIS, F. R., 'Reality and Sincerity' (*Scrutiny*, 1952–3), in id., *The Living Principle* (New York, 1975).

LEONARD, MARY HALL, 'A Problem in Prosody', *Poetry Journal* (April 1917), 14–30.

LEVY, WILLIAM, *William Barnes* (Dorchester, 1960).

LEWIS, C. DAY, 'The Lyrical Poetry of Thomas Hardy', *Proceedings of the British Academy*, 37 (1951), 155–74.

LEWES, GEORGE HENRY, *The Story of Goethe's Life* (London, 1873).

LINTON, W. J. and R. H. STODDARD, *English Verse: Lyrics of the 19th Century* (London, 1884) (Hardy's copy, Dorset County Museum).

LOCKER, FREDERICK, *London Lyrics* (London, 1878; originally pub. in 1857) (Hardy's copy, Cox catalogue).

MACKAIL, J. W. (ed.), *Select Epigrams from the Greek Anthology*, 3rd edn. (London, 1911) (Hardy's copy, Dorset County Museum).

MACKENZIE, NORMAN, 'Hopkins and the Prosody of Sir Thomas Wyatt', *Hopkins Quarterly*, 8 (1982), 63–73.

McCARTHY, LILLAH, *Myself and My Friends* (London, 1933).

McKERROW, R. B., 'The Use of So-Called Classical Metres in Elizabethan Verse', *MLQ* 4 (1901), 172–80; 5 (1902), 6–13.

MALOF, JOSEPH, *A Manual of English Meters* (Westport, Conn., 1970).

MARDON, J. VERA, *Thomas Hardy as Musician*, No. 15 in J. Stevens Cox (ed.), *Monographs on the Life of Thomas Hardy* (Beaminster, 1964).

MARSDEN, KENNETH, *The Poems of Thomas Hardy* (New York, 1969).

MASON, JOHN, *An Essay on the Power and Harmony of Prosaic Numbers* (London, 1749).

—— *An Essay on the Power of Numbers and the Principles of Harmony in Poetical Composition* (1749) (London, 1761).

MASSON, DAVID, 'Milton's Versification and His Place in the History of English Verse', in id. (ed.), *The Poetical Works of John Milton*, vol. i (London, 1874).

MASSON, DAVID I., 'Vowel and Consonant Patterns in Poetry', *JAAC* 12 (1953), 213–27; repr. in S. Chatman and S. Levin (eds.), *Essays on the Language of Literature* (Boston, 1967).

—— 'Word and Sound in Yeats' "Byzantium"', *ELH* 20 (1953), 136–60.

MAYOR, J. B., *Chapters on English Metre* (London, 1886; rev. form of work originally publ. in the *Transactions of the Philological Society*, 1873–7; 2nd rev. edn., Cambridge, 1901).

MILLER, J. HILLIS, *The Linguistic Moment: From Wordsworth to Stevens* (Princeton, 1985).

MILLGATE, MICHAEL, *Thomas Hardy: A Biography* (New York, 1982).

—— *Thomas Hardy: His Career as a Novelist* (London, 1971).

MILNES, RICHARD MONCKTON (LORD HOUGHTON), *The Poetical Works*, 2 vols. (London, 1876) (Hardy's copy, Dorset County Museum).

MITFORD, WILLIAM, *An Essay upon the Harmony of Language* (London, 1774).

'Modern Developements [sic] in Ballad Art', *Edinburgh Review*, 213 (1911), 153–79.

MOORE, T. STURGE, D. S. MACCOLL, *et al.*, 'English Numbers'. Correspondence in *TLS* (9 Jan. 1919), 20–1; (16 Jan.), 33; (23 Jan.), 45; (30 Jan.), 56–7; (6 Feb.), 69; (13 Feb.), 83–4; (20 Feb.), 97; (27 Feb.), 112; (6 Mar.), 125; (13 Mar.), 137; (20 Mar.), 151–2; (27 Mar.), 164–5.

MURRY, JOHN MIDDLETON, 'The Poetry of Thomas Hardy', *Athenaeum* (1919); repr. in id., *Aspects of Literature* (London, 1920).

—— *The Problem of Style* (London, 1922).

NEWBOLT, HENRY, 'A New Study of English Poetry', *English Review*, 10 (1911–12), 285–301, 657–72.

—— (ed.), *An English Anthology of Prose and Poetry* (London, 1921) (Hardy's copy, Mathews catalogue).

—— *The Tide of Time in English Verse* (London, 1925).

NEWMAN, JOHN HENRY, *Verses on Various Occasions* (London, 1868) (Hardy's copy, Dorset County Museum).

ODELL, J., *An Essay on the Elements, Accents, and Prosody of the English Language* (London, 1806).

OMOND, THOMAS STEWART, *English Metrists*, rev. edn. (Oxford, 1921; repr. New York, 1968; earlier edns. London, 1907; Tunbridge Wells, 1903).

—— 'Is Verse a Trammel?', *Gentleman's Magazine* (Jan–June 1875), 344–54.

—— *A Study of Metre* (London, 1903).

—— and T. B. RUDMOSE-BROWN, ' "Inverted Feet" in Verse', *Academy* 75 (1908), 329–30, 351–2, 378–9, 401, 429, 451–2, 475–6, 498–9, 524, 548–9, 571.

OREL, HAROLD, *The Final Years of Thomas Hardy, 1912–1928* (Laurence, Kan., 1976).

OSTRIKER, ALICIA, 'The Three Modes in Tennyson's Prosody', *PMLA* 82 (1967), 273–84.

The Oxford Book of English Verse 1250–1900, ed. Arthur Quiller-Couch (Oxford, 1900) (Hardy's copy, Cox catalogue).

The Oxford Book of Victorian Verse, ed. Arthur Quiller-Couch (Oxford, 1912) (Hardy's copy, Dorset County Museum).

Oxford English Dictionary (Hardy's copy, Cox and Sotheran catalogues).

PALGRAVE, F. T., *The Golden Treasury* (Cambridge, 1861) (Hardy's copy, Dorset County Museum).

PATMORE, COVENTRY, 'The Aesthetics of Gothic Architecture', *British Quarterly Review*, 10 (1849), 46–75.

—— *Amelia, Tamerton, Church-Tower, Etc., with Prefatory Study on English Metrical Law* (London, 1878).

—— *The Angel in the House*, 2 vols. (London, 1860–3; originally publ. in parts, 1854, 1856, 1860, 1862) (Hardy's copy, Dorset County Museum).

—— 'English Metrical Critics', *North British Review*, 27 (1857), 127–61; rev. and repr. with *Amelia*; repr. in various very slightly rev. forms thereafter, including the 1886 *Poems* where it was an appendix entitled 'Essay on English Metrical Law'. The definitive modern edn. is:

—— *Essay on English Metrical Law*, ed. Sister Mary Roth (Washington, DC, 1961); based on the text published with the 1894 *Poems*.

—— *Memoirs and Correspondence*, ed. Basil Champneys, 2 vols. (London, 1900).

—— *Poems*, 4 vols. (London, 1879) (with 'Prefatory Study on English Metrical Law' in vol. i). 2nd collective edn., 2 vols. (London, 1886) (with 'Appendix: Essay on English Metrical Law'). New collective edns. (including the 'Appendix') in 1887, 1890, 1894, 1897, 1900, 1903, 1906.

—— *Principle in Art* (London, 1889).

—— '*In Memoriam*', *North British Review*, 13 (1850), 532–55.

—— '*Maud, and Other Poems*', *Edinburgh Review*, 102 (1855), 498–519.

PATTERSON, CHARLES, 'An Unidentified Criticism by Coleridge Related to Christabel', *PMLA* 67 (1952), 973–88.

PAULIN, TOM, *Thomas Hardy: The Poetry of Perception* (London, 1975).

PERCY, BISHOP THOMAS, *Reliques of Ancient English Poetry*, ed. Robert Aris Willmott (London, 1857; originally pub. 1765) (Hardy's copy, Dorset County Museum).

PERKINS, DAVID, *A History of Modern Poetry* (Cambridge, Mass., 1976).

PERLOFF, MARJORIE, *Rhyme and Meaning in the Poetry of Yeats* (The Hague, 1970).

The Phoenix Nest, ed. Hyder Rollins. (Cambridge, Mass., 1931; originally publ. 1593; new edns. in 1814, 1815, 1867).

PINION, F. B., *A Commentary on the Poems of Thomas Hardy* (London, 1976).

—— *A Hardy Companion: A Guide to the Works of Thomas Hardy and their Background* (London, 1968).

POE, EDGAR ALLEN, 'The Philosophy of Composition' (1846) and 'The Rationale of Verse' (1848, rev. form of 1843 essay), in vol. iii of *Works*, 4 Vols. ed. John Ingram (Edinburgh, 1874–5) (Hardy's copy, Dorset County Museum).

POUND, EZRA, and MARCELLA SPANN, *Confucius to Cummings, An Anthology of Poetry* (New York, 1964).

PURDY, RICHARD LITTLE, *Thomas Hardy: A Bibliographical Study* (London, 1954).

PUTTENHAM, GEORGE, *The Arte of English Poesie* (1589), in G. Gregory Smith (ed.), *Elizabethan Critical Essays*, vol. ii (London, 1904).

PYRE, JAMES F., *The Formation of Tennyson's Style* (Madison, Wis., 1921).

RANKIN, J. W., 'Rime and Reason', *PMLA* 44 (1929), 997–1004.

RANSOM, JOHN CROW, 'Hardy—Old Poet', *New Republic*, 126 (12 May 1952), 16, 30–1.

REES, JOAN, *Dante Gabriel Rossetti: Modes of Self-expression* (Cambridge, 1981).

RICKEY, MARY ELLEN, 'Herbert's Technical Development', *JEGP* 62 (1963), 745–60.

ROE, RICHARD, *The Principles of Rhythm* (Dublin, 1823).

RUSKIN, JOHN, *Elements of English Prosody; for Use in St. George's Schools* (Orpington, 1880).

—— *Works*, ed. E. T. Cook and Alexander Wedderburn, 39 vols. (London, 1903–12).

SAINTSBURY, GEORGE, 'English Versification', introduction to André Loring (Lorin Lathrop) (ed.), *The Rhymer's Lexicon* (London, 1905) (Hardy's copy, Hollings catalogue).

—— (ed.), *French Lyrics* (London, 1883) (Hardy's copy, Dorset County Museum).

—— *A History of Elizabethan Literature* (London, 1887).

—— *A History of English Prose Rhythm* (London, 1912).

—— *A History of English Prosody*, 3 vols. (London, 1906, 1908, 1910).

—— *Historical Manual of English Prosody* (London, 1910).

—— 'The Prosody of the Nineteenth Century', in *The Cambridge History of English Literature*, vol. xiii (Cambridge, 1917).

—— 'Recent Verse', *The Academy* (14 May 1881), 351–3.

SALTER, C. H., *Good Little Thomas Hardy* (Totowa, NJ, 1981).

SAPPHO, *Poems*, trans. Henry Thornton Wharton, 3rd edn. (London, 1895) (Hardy's copy, Dorset County Museum).

SAY, SAMUEL, *Poems on Several Occasions: and Two Critical Essays, viz. The First, on the Harmony, Variety, and Power of Numbers, whether in Prose or Verse. The Second, on the Numbers of 'Paradise Lost'* (London, 1745).

SCHIPPER, JACOB, *Englische Metrik: in Historischer und Systematischer Entwickelung Dargestellt*, 2 parts (Bonn, 1881–8).

—— *A History of English Versification* (Oxford, 1910).

SCHNEIDER, ELIZABETH, *The Dragon in the Gate: Studies in the Poetry of Gerard Manley Hopkins* (Berkeley, 1968).

Sequences from the Sarum Missal, trans. Charles B. Pearson (London, 1871) (Hardy's copy, Purdy, 237; Holmes Listing: Purdy collection).

The Shakespeare Anthology 1592–1616 A.D., ed. Edward Arber (London, 1899) (Hardy's copy, Dorset County Museum).

SHAPIRO, KARL, 'English Prosody and Modern Poetry', *ELH* 14 (1947), 77–92.

—— and ROBERT BEUM, *A Prosody Handbook* (New York, 1965).

SHERIDAN, THOMAS, *Lectures on the Art of Reading*, 2 vols. (London, 1775).

SHERMAN, ELNA, 'Music in Thomas Hardy's Life and Work', *Musical Quarterly*, 26 (1940), 419–45.

—— 'Thomas Hardy: Lyricist, Symphonist', *Music and Letters*, 21 (1940), 143–71.

SIDNEY, SIR PHILIP, 'The Psalmes of David', 1–48, in *Complete Poems*, ed. Alexander Grosart, 3 vols. (London, 1877) (Hardy's copy, Dorset County Museum).

SKEAT, WALTER W., 'An Essay on Alliterative Poetry', in J. W. Hales and F. J. Furnivall (ed.), *Bishop Percy's Folio Manuscript: Ballads and Romances*, 3 vols. (London, 1867–8).

—— 'An Essay on the Language and Versification of Chaucer', in R. Morris (ed.), *The Poetical Works of Geoffrey Chaucer*, 6 vols. (London, 1866).

—— 'Preface' to id. (ed.), *The Lay of Havelok the Dane* (London, 1868).

SNELL, ADA, 'The Meter of Christabel', in *The Fred Newton Scott Anniversary Papers* (Chicago, 1929).

Songs from the Novelists, ed. William Davenport Adams (London, 1885).

SOUTHEY, ROBERT, Preface to *A Vision of Judgment* (1821), in *Poetical Works* (London, 1845) (Hardy's copy, Dorset County Museum).

SOUTHWORTH, JAMES, *The Poetry of Thomas Hardy* (New York, 1947).

STEELE, JOSHUA, *An Essay Towards Establishing the Melody and Measure of Speech* (London, 1775).

STEIN, ARNOLD, 'Structures of Sound in Donne's Verse', *Kenyon Review* 13 (1951), 20–36, 256–78.

STEPHENSON, EDWARD, 'Hopkins' "Sprung Rhythm" and the Rhythm of *Beowulf*', *Victorian Poetry*, 19 (1981), 97–116.

STERNHOLD, THOMAS, and JOHN HOPKINS, *et al.*, *The Whole Book of Psalms* (1562).

STEVENSON, ROBERT LOUIS, 'On Style in Literature: its Technical Elements', *Contemporary Review*, 47 (1885), 548–61; repr. as 'On

Some Technical Elements of Style in Literature', in id., *Essays in the Art of Writing* (London, 1905).

STEWART, GEORGE, 'A Method Toward the Study of Dipodic Verse', *PMLA* 39 (1924), 979–89.

—— *Modern Metrical Technique as Illustrated by Ballad Meter (1700–1900)* (New York, 1922).

STONE, WILLIAM JOHNSON, *On the Use of Classical Metres in English* (London, 1899); rev. and repr. in a combined edn. with Bridges's *Milton's Prosody* in 1901.

SWINBURNE, ALGERNON, *Poems and Ballads* (1866), 5th edn. (London, 1873) (Hardy's copy, Dorset County Museum).

SYMONDS, JOHN ADDINGTON, *Blank Verse* (London, 1895); repr. of 'Appendix' to *Sketches and Studies in Italy* (1879), in turn a repr. of 'Blank Verse', *Cornhill Magazine*, 15 (1867), 620–40, and 'The Blank Verse of Milton', *Fortnightly Review*, 16 (Dec. 1874), 767–81.

TAMKE, SUSAN, *Make a Joyful Noise unto the Lord: Hymns as a Reflection of Victorian Social Attitudes* (Athens, 1978).

TARLINSKAJA, MARINA, *English Verse: Theory and History* (The Hague, 1976).

TATE, NAHUM, and NICHOLAS BRADY, *A New Version of the Psalms of David* (1696), bound with *The Book of Common Prayer* (Cambridge, 1858) (Hardy's copy, Dorset County Museum).

TAYLOR, DENNIS, *Hardy's Poetry, 1860–1928* (London, 1981).

—— 'Victorian Philology and Victorian Poetry', *Victorian Newsletter* 53 (Spring, 1978), 13–16.

—— 'Hardy and Wordsworth', *Victorian Poetry*, 24 (1986), 441–54.

THELWALL, JOHN, *Illustrations of English Rhythmus* (London, 1812).

THOMPSON, JOHN, *The Founding of English Meter* (New York, 1961).

THOMSON, JAMES, JAMES BEATTIE, *et al.*, *Poetical Works* (London, 1863) (Hardy's copy, Dorset County Museum).

TOLMAN, ALBERT, 'The Symbolic Value of English Sounds', in id., *The Views About Hamlet and Other Essays* (Boston, 1904); originally in shorter form in *The Atlantic Monthly* 75 (1895), 478–85.

TYRWHITT, THOMAS, *An Essay on the Language and Versification of Chaucer*, vol. iv of id. (ed.), *The Canterbury Tales of Chaucer* (London, 1775).

UDAL, JOHN, *Dorsetshire Folk-Lore* (Hertford, 1922).

ULLMANN, STEPHEN, *Semantics* (New York, 1962).

VAN DOREN, MARK, *Autobiography* (New York, 1958).

—— 'Lyrics and Magic', *Nation* (New York), 116 (31 Jan. 1923), 125.

VAUGHAN, HENRY, *Sacred Poems and Pious Ejaculations* (London, 1897) (Hardy's copy, Dorset County Museum).

WALKER, JOHN, *A Rhyming Dictionary of the English Language*, rev. with a preface by John Longmuir (London, 1865) (Hardy's copy, Dorset County Museum).

WARD, THOMAS H. (ed.), *The English Poets*, 5 vols. (London, 1895–1918; first pub. in 4 vols. in 1880) (Hardy's copy, Dorset County Museum).

WATTS, THEODORE, 'Poetry', in *Encyclopaedia Britannica*, 9th edn.

WEBB, DANIEL, *Observations on the Correspondence Between Poetry and Music* (London, 1769).

—— *Remarks on the Beauties of Poetry* (London, 1762).

WEEKS, L. T. 'The Order of Rimes of the English Sonnet', *MLN* 25 (1910), 176–80.

WELLEK, RENÉ, *A History of Modern Criticism: 1750–1950*, vol. ii, *The Romantic Age* (New Haven, 1955).

—— and AUSTIN WARREN, *Theory of Literature*, 2nd edn. (New York, 1956).

WESLING, DONALD, 'The Prosodies of Free Verse', in Reuben Brower (ed.), *Twentieth Century Literature* (Cambridge, Mass. 1971).

WHALLEY, GEORGE, 'Coleridge on Classical Prosody: An Unidentified Review of 1797', *Review of English Studies*, NS 2 (1951), 238–47.

WHITE, GLEESON (ed.), *Ballades and Rondeaus, Chants Royals, Sestinas, Villanelles, etc.* (London, 1887) (Hardy's copy, Holmes Listing: Purdy collection).

WILSON, CARROLL, *Thirteen Author Collections of the Nineteenth Century* (New York, 1950).

WIMSATT, WILLIAM, 'In Search of Verbal Mimesis', *Yale French Studies*, 52 (1975), 229–48.

—— 'The Rule and the Norm: Halle and Keyser on Chaucer's Meter', in S. Chatman (ed.) *Literary Style: A Symposium* (London, 1971); originally in *College English*, 31 (1970), 774–88.

—— and MONROE BEARDSLEY, 'The Concept of Meter: An Exercise in Abstraction', in Wimsatt, *Hateful Contraries: Studies in Literature and Criticism* (Louisville, Ky., 1965); originally in *PMLA* 74 (1959): 585–98.

WOODRING, CARL, 'Onomatopoeia and Other Sounds in Poetry', *College English*, 14 (1953), 206–16.

WRIGHT, WALTER, *The Shaping of 'The Dynasts'* (Lincoln, Nebr., 1967).

YEATS, WILLIAM BUTLER, 'A General Introduction for My Work' (1937), in id., *Essays and Introductions* (New York, 1961).

Catalogues

COLBY COLLEGE LIBRARY, Catalogue of the Thomas Hardy collection. Available from Special Collections, Colby College, Waterville, Me.

COX, J. STEVENS (ed.), *The Library of Thomas Hardy*: repr. of the Hodgson 1938 Sales Catalogue, in *Monographs on the Life, Times and Works of Thomas Hardy*, No. 52 (Guernsey, 1969).

DORSET COUNTY MUSEUM, 'Catalogue of the Thomas Hardy Library', compiled by Shigeru Fujita, with additions by Dennis Taylor: available from Special Collections, Colby College Library, Waterville, Me.

EXPORT BOOK COMPANY, *Selections from the Library of . . . Thomas Hardy*, Sales catalogue No. 287 (Preston, 1938).

Grolier Club Centenary Exhibition of the Works of Thomas Hardy: A Descriptive Catalogue (Waterville, Me., 1940).

HOLLINGS, FRANK, *Modern Times . . . With . . . Books from the Library of Thomas Hardy*, Sales Catalogue No. 212 (London, n.d.).

HOLMES, DAVID, *English First Editions . . . Books from the Library of Thomas Hardy*, Sales Catalogue No. 5 (Boston, n.d.).

—— Listing of books owned by Thomas Hardy; unpub. manuscript.

MAGGS BROS., *Thomas Hardy: A Collection of Books from his Library at Max Gate, Dorchester*, Sales Catalogue No. 664 (London, 1938).

MATHEWS, ELKIN, *Books from the Library of Thomas Hardy*, Sales Catalogue 77 (London, 1939).

SOTHERAN, HENRY, *A Catalogue of General Literature, Including Thomas Hardy's Annotated Copy of the Oxford English Dictionary*, Sales Catalogue No. 967 (London, 1979).

UNIVERSITY OF TEXAS, Austin, Texas, Humanities Research Center Collection of Thomas Hardy's Books and Manuscripts, Catalogue.

WREDEN, WILLIAM, *A Selection of Books from the Library of Thomas Hardy*, Sales Catalogue No. 11 (Burlingame, Calif., 1938).

Recordings

BURTON, RICHARD, *The Poetry of Thomas Hardy*, Caedmon Records, TC1140, 1960.

HOLM, IAN, *et al.*, *Thomas Hardy*, Decca Records, Argo, PLP 1053, 1968.

THOMAS, DYLAN, 'An Introduction to Thomas Hardy', in *An Evening with Dylan Thomas*, 10 Apr. 1950, Univ. of California. Caedmon Records, TC1157, 1963.

INDEX